Taking a Stand: Cultivating a New Relationship with the World's Forests

JANET N. ABRAMOVITZ

Ashley T. Mattoon, *Staff Researcher*

Jane A. Peterson, *Editor*

WORLDWATCH PAPER 140
April 1998

FINANCIAL SUPPORT for the Institute is provided by the Geraldine R. Dodge Foundation, the Ford Foundation, the Foundation for Ecology and Development, the William and Flora Hewlett Foundation, W. Alton Jones Foundation, John D. and Catherine T. MacArthur Foundation, Charles Stewart Mott Foundation, the Curtis and Edith Munson Foundation, David and Lucille Packard Foundation, Rasmussen Foundation, Rockefeller Brothers Fund, Rockefeller Financial Services, Summit Foundation, Surdna Foundation, Turner Foundation, U.N. Population Fund, Wallace Genetic Foundation, Wallace Global Fund, Weeden Foundation, and the Winslow Foundation. The Institute also receives financial support from the Friends of Worldwatch and from our Council of Sponsor members: Toshishige Kurosawa, Kazuhiko Nishi, Roger and Vicki Sant, Robert Wallace, and Eckart Wintzen.

Table of Contents

The views expressed are those of the author and do not necessarily
represent those of the Worldwatch Institute; of its directors, officers, or
staff; or of its funding organizations.

ACKNOWLEDGMENTS: I would like to express my appreciation to a number of individuals for their contributions. Thanks to the following reviewers for their invaluable comments on drafts of this paper: Susan Braatz, Dirk Bryant, Dominick DellaSala, Nigel Dudley, Nels Johnson, William Mankin, Jeff McNeely, Norman Myers, Scott Paul, Nigel Sizer, and Tim Wilson. Thanks to colleagues at Worldwatch: Chris Flavin and Hilary French for review; Payal Sampat for proofreading; Lori Brown, our librarian; Liz Doherty for layout and design; and Dick Bell, Mary Caron, and Amy Warehime of the communications department for coordinating production and outreach. Many thanks to Jane Peterson for her expert editing. Finally, special thanks to Ashley Mattoon for her outstanding research assistance and enthusiasm.

JANET ABRAMOVITZ is a Senior Researcher at the Worldwatch Institute. She leads the Institute's Natural and Social Systems Research Team and conducts her own research on biodiversity, natural resources, and social equity as well. She is a co-author of the Institute's yearly reports, *State of the World* and *Vital Signs*. She has written about ecosystem services in the *State of the World 1997* chapter, "Valuing Nature's Services." Her last Worldwatch Paper was *Imperiled Waters, Impoverished Future: The Decline of Freshwater Ecosystems*. Prior to joining Worldwatch she had 15 years of experience working at World Resources Institute, the National Park Service, and other governmental and nongovernmental organizations.

Introduction

For millennia, humanity has left its mark on the world's forests, although much of it was hard to see. During this century, however, the scale and impact of our footprint has expanded, and the forests that we have always pictured as endless are rapidly shrinking before our eyes. Almost half of those once covering the earth are gone, and deforestation is expanding and accelerating. Between 1980 and 1995 alone at least 200 million hectares of forests vanished—an area larger than Mexico. The deteriorating health and quality of much of the remaining forests weaken their ability to support the species and services they once did, rendering them vulnerable to further decline.[1]

Whereas deforestation was once a localized problem, now it affects the entire planet. The blinding smoke created by the fires that swept through the forests of Indonesia in 1997–98 helped bring the global dimension of catastrophic forest loss into focus when it darkened neighboring countries. The growing understanding of deforestation's role in altering the earth's climate also increases awareness of its global impact.[2]

We have long viewed forests as wasted or even hostile space, valuable only if the trees were cut for timber and the land was tamed for agriculture or settlements. To be sure, there are some early and ongoing examples of sound forest management based on an understanding of forests as vital ecosystems. But destructive practices reflect the more common relationship with forests, one that continues today. In ancient times, the Mediterranean region was moister and

more wooded before trees were cut for shipping, fuel, building, and agriculture. The Philippines' loss of 90 percent of its
primary forest during the timber boom of the 1970s provides
a more recent example of humankind's short-sightedness.[3]

Of course, change now happens much faster and on a
much larger scale than it did in previous generations, as the
Philippines experience illustrates. People no longer log with
an ax or hand saw and a horse, but with chain saws, tractors,
helicopters, and mechanical harvesters. Fires have been used
as a tool for clearing trees and undergrowth for many cen-
turies. But today's fires are no longer restricted to a hectare
or two; often they are used to raze hundreds or thousands of
hectares at a time.

The people clearing vast tracts today are generally the
ones engaged in maximizing short-term profits through
wholesale conversion to satisfy distant markets rather than
those who know the land and have a long-term commitment to it. Indeed, the gap between the people who profit
from deforestation and those who shoulder the loss yawns
ever wider. In Indonesia, commercial outfits (many with personal and financial ties to the president) deliberately set fires
that burned 2 million hectares of forest in 1997 to clear land
for palm oil, pulp, and rice plantations or to cover the tracks
of illegal logging. Haze shrouded much of Indonesia, as well
as neighboring Malaysia, Singapore, Brunei, southern
Thailand, and the Philippines, forcing over 20 million people to breathe hazardous air for months.[4]

When forests disappear, we lose much more than just
timber, we also lose many critical goods and services. The top
150 non-wood forest products traded internationally are
worth over $11 billion per year (not counting the even greater
local value of these products) and employ millions of people.
Forests shelter countless species and provide habitat for other
useful organisms that pollinate crops and control disease-carrying pests. And without forest cover to protect a watershed,
tens of tons of soil can wash away from each hectare of
denuded land, reducing the reliability of water supplies.
Flooding and drought become more extreme. Recent defor-

estation in India's Ganges River valley has caused heavy flooding and property damage of $1 billion per year. The value of each kilometer of mangrove forests in Malaysia for coastal flood control alone is calculated at $300,000.[5]

The world's forests now lose more carbon to the atmosphere than they absorb—a recent and radical shift in the functioning of these critical ecosystems—fueling global climate change. One quarter of all the atmospheric carbon produced by human activities comes from cutting and burning forests. Those that burned in Indonesia sent as much carbon into the atmosphere in a few months as all of Europe's industrial activity did in a year.[6]

Much forest exploitation and conversion is encouraged and subsidized by governments. In both developed and developing countries, these policies result in forests being sold at prices far below what the timber alone is worth. In Belize, a Malaysian logging firm paid about $1 a hectare for timber rights. Indonesia's give-away timber concessions cost the government $2.5 billion in lost revenues in 1990 alone. In the United States, timber sales from national forests lost over $1 billion from 1992 to 1994. Timber supply remains the primary objective of U.S. national forest management despite the fact that recreational use adds 28 times as much money and 34 times as many jobs to the economy. Subsidies for below-cost logging, processing, roads, and infrastructure are so large that governments are essentially paying private interests to take the timber and convert the land to other uses.[7]

Globalization and the lowering of barriers to international trade and investment allow corporations to roam the world seeking more profitable forest opportunities. Legal international trade in forest products has almost tripled since 1970. Meanwhile, illegal and unrecorded trade has likely expanded far faster, encouraged by lax enforcement of domestic forest laws and gaps in international ones. The Brazilian government, for example, estimates that 80 percent of the timber harvest in the Amazon violates the law.[8]

A major force driving these changes is the growing demand for forest products, stoked mainly by rising affluence. Since 1961, consumption of wood has doubled and paper use has more than tripled. Today, less than one fifth of the world's population living in Europe, the United States, and Japan consumes over half the world's industrial timber and more than two thirds of its paper. Japan alone consumes almost as much paper as the entire nation of China, a country with nearly 10 times as many people. In the next 15 years, global demand for paper is expected to grow by half again as industrial countries continue to raise their already high levels of consumption and developing countries try to catch up.[9]

More and more wood is going into disposable products intended for a one-time use, like paper and disposable chopsticks. Today, about 40 percent of the world's industrial timber harvest ends up in paper. In the United States, nearly one fifth of all lumber is used to make shipping pallets, most of which are quickly discarded. In the United Kingdom, 130 million trees' worth of paper is discarded after use each year.[10]

The good news is that innovative ways of satisfying the need for forest products less wastefully are also being pursued, from reducing waste in the forest and in processing, to more efficient use of building material, to expanding paper recycling. Over 40 percent of the world's paper is now recovered and recycled; in many countries the portion is even higher. Recycling is so successful that today over one third of the fiber used to make paper comes from recovered paper, up from just under one quarter in 1970. But, recycling alone won't meet the rising demand for paper. Reducing wasteful consumption is also required.[11]

A shift to more sustainable forest management for a wider range of goods and services is beginning. People are starting to understand that forests can be managed both for timber *and* for the services people need—watershed protection, habitat, climate and water regulation, and so forth. New tools for calculating the other values of a forest, and evolving forest management practices are enabling this shift, and a diverse constituency—scientists, environmentalists, local people, business and government leaders—is demanding it. Large forest product companies as well as small landowners are adopting better management practices. A credible system of certifying products from well-managed forests is now in place, and consumers are starting to express their preference for these products.

Clearly, sustaining forests for the next century and beyond calls for cultivating a new relationship *with* forests *in* the forests. It also requires reforming domestic policies and pricing, reducing waste and overconsumption, recognizing communities' rights to their forests, and ensuring that laws are strengthened and enforced. International mechanisms like the Framework Convention on Climate Change and the Convention on Biological Diversity offer opportunities for international cooperation as does strengthening forest monitoring. There is considerable room for improvement in all these areas that can benefit forests and quality of life while also boosting national and local economies.

Sustaining forests for future generations will mean recognizing that their real wealth lies in their healthy ecosystems—and appreciating how much we depend upon them. The loss of these ecosystems is no longer just a local problem. The scale and consequences of their decline reveal that we are all members of a threatened forest community—a global community challenged to cultivate a new relationship with the forests.

What Forests Do

Although it is often assumed that the greatest value can be extracted from a forest by maximizing timber and pulp production or converting it to agriculture, other uses generally regarded as free or simply not noticed are often more valuable. These uses can also be sustained over the long term and benefit more people. For instance, when alternative management strategies for the mangrove forests of Indonesia's Bintuni Bay were compared, taking into account the value of fish, locally used products, and erosion control, the most profitable strategy was to keep the forest standing, yielding $4,800 per hectare. Cutting the timber, on the other hand, netted only $3,600 per hectare. Not logging the forest would also allow continued local uses of the area worth $10 million a year, providing 70 percent of local income, and would protect fisheries worth $25 million a year.[12]

Forests provide many products and services beyond timber. They produce non-wood materials such as food, fodder, fish, oils, resins, spices, fragrances, and medicines. Among their services are purifying and regulating water supplies; absorbing and decomposing wastes; cycling nutrients; creating and maintaining soils; providing pollination, pest control, habitat, and refuge; moderating disturbances such as floods and storms; and regulating local and global climates. Forests also give educational, recreational, aesthetic, and cultural benefits. Finally, they yield sustenance and livelihoods for hundreds of millions of people, including those who are excluded from the formal economy.[13]

Non-wood forest products are a significant source of revenue worldwide. FAO estimates that over 150 non-wood forest products (NWFP) are traded internationally in significant amounts, altogether worth $11.1 billion a year. (By comparison, the trade in wood products like timber, pulp, and paper is worth about $142 billion a year.) Rattan, for example—a vine that grows naturally in tropical forests—is widely used to make furniture. Global trade in rattan generates $2.7 billion in exports annually, and in Asia it employs a half-million people. In Thailand, the value of rattan exports is equal to 80 percent of legal timber exports. In India, such "minor" products account for three fourths of the net export earnings from forest produce, and provide more than half of formal employment in the forestry sector. And in Indonesia, hundreds of thousands of people make their livelihoods collecting and processing NWFPs for export, a trade worth at least $25 million a year. Many of the forests that sustained these products were destroyed in the 1997–98 fires.[14]

Perhaps even more valuable than these global economic contributions is the role of harvested wild forest goods in local economies and households, where they are part of flexible and sustainable livelihood systems. Wild products collected for personal consumption and traded in markets include vegetables, fruits, honey, game meat, fish, medicines, dyes, and materials for thatching, cording, and weaving. In rural Laos, at least 141 forest products are gathered, mainly by women, for sale or home use. In Ghana, most people depend on wildlife for most of their protein. Access to wild resources can mean survival during times of famine and in the gap before the next agricultural harvest.[15]

NWFPs are also important in industrial nations. Cork, for example, a bark from the cork oak tree used to make stoppers, flooring, engine gaskets, and other products, generates over $300 million a year and employs about 14,000 workers in Portugal, the world's largest producer nation. In Canada and the United States, the economic contribution of NWFPs such as mushrooms and Christmas greens is now being recognized.[16]

Access to wild products is especially important to people who are usually excluded from more formal employment. Most workers in the world's formal and informal NWFP economy are women. Even among the renowned Amazonian rubber tappers, women are responsible for the processing that adds significant value to the products. Because a large share of women's income goes to support their families' health and welfare, this undercounted economy makes a substantial contribution. In India, NWFPs meet the household needs of tribal and forest communities and the landless, and the collection, processing, and sale of these products are their most important source of income. In fact, Indian biologist Madhav Gadgil notes that "a third or more of Indian people behave as 'ecosystem people'—households whose quality of life is intimately linked to the productivity and diversity of living organisms in their own [areas]."[17]

Producing non-wood commodities is only part of what forests do. They also provide habitat for bees and other pol-

linators and for birds that control disease-carrying and agricultural pests. And they are home to a majority of the
world's known terrestrial species as well as bountiful aquatic life contained in the many rivers running through them.
(At least one third of the world's 9,000 known fish species
live in the Amazon River and its naturally flooded forest.)
The forest canopies break the force of the winds and reduce
rainfall's erosive impact on the ground. Their roots hold soil
in place, further stemming erosion. In purely monetary
terms, a forest's watershed protection value alone can exceed
the value of its timber. Forests also act as effective water-
pumping and recycling "machinery," helping to stabilize
local climates. Through photosynthesis, they generate oxygen while absorbing and storing carbon, making them
essential to the stability of climate worldwide.[18]

Beyond these general functions, certain services are
specific to particular kinds of forests. Mangrove forests and
coastal wetlands, for example, buffer coasts from storms and
erosion, cycle nutrients, serve as nurseries for coastal and
marine fisheries, and supply critical resources to local communities. Coastal protection will be especially important as
climate change whips up more violent and unpredictable
storms. And the peat swamp forests of Malaysia and
Indonesia play a vital role in water storage and supply for
human consumption and agricultural use.[19]

The planet's water moves in a continuous cycle, falling
as precipitation and flowing slowly over the landscape to
streams and rivers and ultimately to the sea, being absorbed
and recycled by plants along the way. Yet human actions
have changed even that most fundamental force of nature
by removing natural plant cover, draining swamps and wetlands, separating rivers from their floodplains, and paving
over the land. The slow natural movement of water across
the landscape is vital for refilling nature's underground
reservoirs, or aquifers, from which we draw much of our
water. In many places, water now races across the land much
too quickly, causing flooding at some times and droughts at
others, while failing to recharge water supplies.

The value of a forested watershed lies in its capacity to absorb and cleanse water, recycle excess nutrients, hold soil in place, and prevent flooding. When plant cover is removed or disturbed, water and wind also carry valuable topsoil with them as they rush over the land. Extreme soil erosion can jeopardize downhill communities, the productivity of rivers and coasts, and the operation of dams and navigation. According to agricultural ecologist David Pimentel, erosion rates on exposed soil—such as that on deforested lands—can be many thousand times higher than natural rates. Under natural conditions, each hectare of land loses somewhere between 0.004 and 0.05 tons of soil to erosion each year— which is more than replaced by natural soil-building processes. On lands that have been logged or converted to agriculture, however, erosion typically washes away 17 tons per hectare a year in the United States or Europe, and 30 to 40 tons in Asia, Africa, or South America. On severely degraded land, erosion can rise to 100 tons per year. The eroded soil transports nutrients and sediments valuable to the system it leaves, but often harmful to its ultimate destination. The erosion, landslides, sedimentation, and higher water temperatures caused by logging can render streams unfit for aquatic life for decades after logging has ended.[20]

The costs of losing forests' services can illustrate just how valuable these "free" services really are. Deforestation in India's Ganges River valley has caused heavier flooding and property damage of $1 billion per year. According to 1988 government estimates, soil losses due to deforestation cost the Canadian province of British Columbia $80 million a year. They projected that losses to wood productivity alone would increase by $10 million per year. In the U.S. Pacific Northwest, where many hundreds of landslides now occur annually, a study found that 94 percent originated from clear-cuts and logging roads. The torrents of water and debris from degraded watersheds caused billions of dollars in damage in 1996 alone.[21]

Forests' regulation of the global climate is one important service that has drawn attention in recent years. One of the planet's first ecosystem services was the production of oxygen over billions of years of photosynthetic activity, which allowed oxygen-breathing organisms—such as human beings—to evolve. Since the industrial revolution, however, humans have begun to unbalance the global climate regulation system by generating too much carbon dioxide through burning fossil fuels and reducing forest area, leaving less to soak up the carbon that we emit. Cutting and burning forests and peat deposits only makes the problem worse by adding still more carbon to the atmosphere. The cumulative effects of such local land use changes have global implications. The 1997 fires in Asia alone released about as much carbon into the atmosphere as all of the factories, power plants, and vehicles in Western Europe do in a year. Climate change will further trigger a host of serious ecological disruptions in the world's forests which will further undermine their ability to support life and regulate the climate.[22]

Without human intervention, forests would be able to store more carbon than they release. But scientists now estimate that during the 1980s there was a net release of carbon into the atmosphere from the world's forests, largely due to deforestation in the tropics. In fact, deforestation contributes about 25 percent of all anthropogenic carbon; the other 75 percent comes from fossil fuel burning.[23]

To give a sense of just how much carbon forests can hold, a mature oak can contain about three tons. A hectare of plantation forest can hold around 200 tons, the same area of mature tropical moist forest can retain over 300 tons, and a hectare of mature Douglas fir forest in the Pacific Northwest can hold over 600 tons. The 737 million acres of forestland in the U.S. store about 60 billion tons of carbon— 40 times the nation's annual carbon emissions.[24]

Because of forests' ability to store vast amounts of carbon in their woody biomass as well as in the peat and organic matter in their soils, there is real potential for forests once again to become a net sink of carbon rather than a source. The New Zealand Forest Research Institute calculated that if that nation planted about 100,000 hectares of plantations a year between 2008 and 2026, the forests could absorb more carbon than the nation will emit from fossil fuel burning. In fact, the Intergovernmental Panel on Climate Change (IPCC) has estimated that over the next 50 years the world's forests could absorb 12–15 percent of the carbon emitted from fossil fuel burning during that time, if steps were taken to restore forests' role as carbon sinks.[25]

Ways to help forests resume this role include reducing deforestation, regenerating degraded forests, establishing forest plantations, employing less disruptive harvest and land use practices, and using the wood that is harvested to make longer-lasting products so the carbon storage is prolonged. For example, wood used for houses and furniture ties up carbon longer than wood turned into paper or fuel. The IPCC has suggested that the conservation and regeneration of tropical forests has the greatest potential to soak up the largest amount of carbon in the future. That is because these forests grow quickly and because tropical deforestation is now a major source of atmospheric carbon.[26]

There is real potential for forests once again to become a net sink of carbon rather than a source.

A number of innovative projects have begun to take advantage of the carbon storage potential of forest ecosystems. Some energy companies and manufacturers, for example, are sponsoring tree planting or forest conservation projects as ways to offset their industrial carbon emissions. Reducing the impact of logging can also lower the amount of carbon released during those activities. Economists estimate the value of standing forest ecosystems for carbon sequestration at several hundred to several thousand dollars per hectare. Conservation of existing forests is likely to be the most cost effective approach and more successful in the long run. As the climate changes, the value of natural systems for regulating climates will only increase.[27]

A recent landmark study helps illuminate the importance of nature's services in supporting human economies by giving the first overall estimate of the current economic value of the world's ecosystem services and natural capital. Robert Costanza of the University of Maryland and colleagues from around the world synthesized the findings of more than 100 studies to compute the value of most of the services that the world's major ecosystem types provide. They calculated that the current economic value of the world's ecosystem services is at least $16–54 trillion per year, perhaps exceeding the gross world product of $28 trillion (in 1995 dollars). If every service for each ecosystem type were measured, the figure would be even higher. Fixing a more accurate price for the benefits from forests is essential, but so too is acknowledging that not everything has a price. Much of a forest's value is quite literally beyond measure.[28]

Assessing Forest Area and Quality

Today, forests cover more than one quarter of the world's total land area (excluding Antarctica and Greenland). Slightly more than half of the world's forests are in the tropics; the rest lie in temperate and boreal (northern) zones. Seven countries hold more than 60 percent of the world's forests: in order of forest area, they are Russia, Brazil, Canada, the United States, China, Indonesia, and the Democratic Republic of Congo (formerly Zaire).[29]

The world's forest estate has declined significantly in both area and quality in recent decades. Almost half—3 billion hectares—of the forests that once blanketed half the earth are gone (Table 1). Most of the loss has occurred in this century through land clearing for timber or for conversion to other uses, such as cattle ranches, agricultural plantations and croplands, or infrastructure development. The area of forest lost just between 1980 and 1995 was at least 200 million hectares—an area larger than Mexico. During the first half of this decade, at least 107 countries experienced a net loss of forest cover. Each year at least another 16 million hectares of natural forest—an area the size of Washington state or more than twice the size of Ireland—is lost in developing countries.[30]

The extent of forest loss and fragmentation was made clear in a recent study by the World Resources Institute (WRI) that identified what it calls "frontier forests"—areas of "large, ecologically intact, and relatively undisturbed natural forests"—which are irreplaceable. The study found that only 22 percent of the world's original forest cover (or 40 percent of current forest cover) remains in these large expanses, about evenly divided between boreal and tropical forest. (See Table 1.) More than 75 percent of the frontier forest is in three large areas: the boreal forests of Canada and Alaska, the boreal forests of Russia, and the tropical forests of the northwestern Amazon Basin and the Guyana shield (Guyana, Suriname, French Guiana, northeastern Brazil, Venezuela, and Colombia). (See Figure 1.) Only 3 percent of the frontier forests lie entirely within temperate zones, mainly in Chile and Argentina.[31]

Only eight countries (Brazil, Suriname, Guyana, Canada, Columbia, Venezuela, Russia, and French Guiana) have large portions of their original forests in vast undisturbed blocks. Seventy-six countries have no frontier forest remaining; 11 others are about to lose the rest of theirs.[32]

Until recently, most forest loss occurred in Europe, North Africa, the Middle East, and temperate North America. By the early part of the twentieth century these regions had been largely stripped of their original cover. Now forest cover in Europe and the United States is gradually increasing, as secondary forests and tree plantations fill in. In the last 30–40 years, most deforestation has occurred in the tropics, where the pace has been accelerating. Indeed, between 1960 and 1990, one fifth of all tropical forest cover disappeared. Asia lost one third, and Africa and Latin America lost about 18 percent each. These regions have continued to lose large portions of forest cover during this decade.[33]

TABLE 1

Forest Cover and Frontiers, by Region

Region	Original Forest	Total Remaining Forest (Frontier and Non-Frontier[1])	Net Annual Change 1990–95 (percent per year)	Total Remaining as Share of Original Forest (percent)	Frontier Forest as Share of Total Remaining Forest (percent)
	(thousand square kilometers[1])				
Africa	6,799	2,302	-0.7	34	23
Asia	15,132	4,275	-0.7	28	20
North & Central America	12,656	9,453	-0.1	75	41
North America	10,877	8,483	0.2	78	44
Central America	1,779	970	-1.2	55	18
South America	9,736	6,800	-0.5	70	65
Europe and Russia	16,449	9,604	—	58	36
Europe	4,690	1,521	0.3	32	1
Russia	11,759	8,083	0.1	69	43
Oceania[2]	1,431	929	-0.1	65	34
WORLD	62,203	33,363	-0.3	54	40

[1]One square kilometer equals 100 hectares. [2]Oceania consists of Papua New Guinea, Australia, and New Zealand.
Source: See endnote 30.

Broad regional overviews such as these can mask the even more severe forest loss that is taking place in some countries and forest types. Fifty-five percent of the total forest loss between 1980 and 1995 took place in just seven countries: Brazil, Indonesia, the Democratic Republic of Congo, Bolivia, Mexico, Venezuela, and Malaysia. Tropical dry forest types, mangrove forests of tropical coasts, and the temperate rain forests of North America have also experienced very high losses.[34]

Deforestation is not the only threat. Serious declines in forest quality are affecting many of the world's forests. While many people in northern countries look at tropical forests with concern, they may be unaware that the temperate forests in their own backyards are the most fragmented and disturbed of all forest types. For example, 95–98 percent of forests in the continental United States have been logged at least once since settlement by Europeans and much of that land has been converted for agriculture or grazing. In Canada, Alaska, and Russia, logging of old growth forest continues. And in Europe, two thirds of the forest cover is gone, while less than 1 percent of old growth remains.[35]

The secondary forests and plantations that are filling in the cut-over forests and abandoned fields are a very different type than the original. The mix of tree and other species has changed, and the age is often more uniform. In addition, the forests are highly manipulated and fragmented. These altered ecosystems usually cannot support the full array of native species and ecological processes and services that characterize natural forests. Many nonnative species—from trees to vines to insect and animal pests—have invaded these woodlands, further disrupting natural ecosystems.[36]

Tree plantations now occupy substantial areas of forestland. Worldwide, there are about 180 million hectares of tree plantations, split about equally between developed and developing countries. Over half of the area is dedicated to intensive production of wood fiber for paper, wood panels, and sawlogs. The remainder have been established on degraded lands for soil and water conservation, carbon offsets, or to

FIGURE 1

Forest Area, by Region, 1996

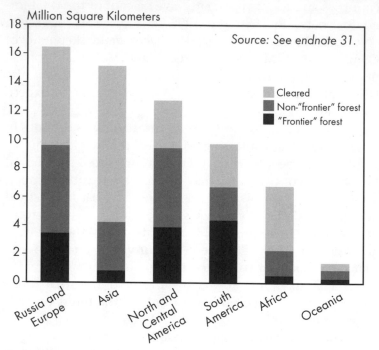

Source: See endnote 31.

Cleared
Non-"frontier" forest
"Frontier" forest

provide local benefits like fuelwood, forage, and employment. In the last 15 years, the area covered by tree plantations
has doubled globally, and it is predicted to increase apace.[37]

Most trees on plantations are grown like crops, often a
single species with a single purpose, grown and harvested in
short periods of time. And many plantations raise non-
native species. In many European countries, the majority of
forest cover is now non-native species. Nearly all of the
forests covering 15 percent of Scotland, for example, are
plantations of non-native conifers. Over 17 percent of
Chile's forests are plantations, primarily Monterey pine
(*Pinus radiata*) from California. New Zealand's 1.6 million
hectares of plantation forest (primarily Monterey pine) are
expanding by 70,000 hectares a year. In some cases, plantations are established not as a way of reforesting degraded

land, but by cutting down intact forest. If this happens, or if people who depended on the natural forest or agricultural land are displaced, one potential benefit of plantations—reducing pressure on natural forests—may not occur.[38]

Atmospheric pollution is also taking a toll on forest quality. Exposure to pollution weakens trees and makes them more vulnerable to pests, diseases, drought, and nutrient deficiencies. This is especially evident in Europe, North America, Asia, and cities throughout the world. More than a quarter of Europe's trees show moderate to severe defoliation from these stresses, according to regular surveys by the United Nations Economic Commission for Europe. The trees in the industrial belt of Poland, the Czech Republic, and what was East Germany have been especially hard hit.[39]

As troubling as the statistics on forest loss and declining quality are, the true picture of the global forest situation is undoubtedly much worse. A major obstacle to assessing the health and extent of the world's forests is the quality of the data assembled by the U.N. Food and Agriculture Organization (FAO), the most widely used source, and the only comprehensive source that spans decades. Data are published years after collection, and pressures and threats are not monitored at all. And since FAO relies largely on self-reporting by governments, and many countries do not have the capacity to carry out systematic forest assessments, the data are of uneven accuracy. Currently there is no global system of independent monitoring in place—either from satellites or on the ground. For its year 2000 Forest Resources Assessment, FAO anticipates using satellite imagery to assess temperate and boreal forests and would like to expand such coverage in tropical countries.[40]

FAO also uses some definitions that can lead to misleading conclusions. For example, "deforestation" occurs when tree cover in an area has been reduced to less than 20 percent in developed countries or less than 10 percent in developing countries. Further, deforestation is defined by FAO as the conversion of forests to other uses such as cropland and shifting cultivation. Forests that have been converted to tree plantations are not counted as deforested, nor are forests intensively logged and left to regenerate. Thus, some of the land reported by countries as forested actually may have no trees on it at all. Using these definitions, 80–90 percent of forest cover can be removed by logging without "deforesting" an area. (Then when small-scale farmers remove the relatively small number of trees remaining, they have, according to the official definition, "deforested" the land. This is why "slash-and-burn" farmers are often blamed for deforestation for which they are not responsible.)[41]

The massive fires in the forests of Indonesia and Brazil in 1997 focused attention on the inadequacy of existing official monitoring. Because the Indonesian government did not have the monitoring capacity, non-governmental organizations (NGOs) and academic institutions collected satellite data and digitized forest concession maps and the locations of fires to prepare detailed maps daily for government officials on the extent and location of fires. The government has since begun to adopt this type of monitoring. Likewise, NGOs took the lead in analyzing the data and reporting the fires in Brazil. In early 1998, the Brazilian government finally released the satellite data it had collected, which showed that the annual area deforested in its Amazon rain forest tripled between 1990–91 and 1994–95 to more than 2.9 million hectares a year. Deforestation was even more extensive than it had been in the late 1980s, when burning in the Amazon sparked international alarm.[42]

Rising Pressures on Forests

In recent decades, the pressures on the world's shrinking forests have intensified. Growing appetites for forest products and forest and agricultural trade are major forces driving the logging and conversion. The output of forest products has accelerated in recent decades. Global production of roundwood—the logs cut for industrial lumber and paper

products or used for fuelwood and charcoal—has risen by
almost 50 percent since 1965. (See Table 2.) The United
States, China, India, Brazil, and Canada combined produce
just over half of all the wood that is used in the world.[43]

According to FAO statistics, about half of the wood cut
worldwide is used for fuelwood and charcoal. Most fuelwood
use is in developing countries, where consumption of this
energy source has kept pace with population growth. (See
Table 3.) (One notable exception is the United States, where
fuelwood production has increased more than fivefold since
1970.) In some developing nations, especially in dry areas like
India and Nigeria, the majority of trees cut are for fuelwood.
But in moist tropical nations such as Malaysia, the vast majority of trees felled are for industrial timber.[44]

In developing countries, most of the live trees that are
cut for fuel are used to make charcoal or in other industrial
applications, such as brick-making and tobacco-curing, or
are burned by city-dwellers. This commercial fuelwood collection, especially when concentrated near cities, can cause
significant local deforestation. Tobacco growers in Brazil
(the world's largest tobacco exporter), for example, use

TABLE 2

Production of Wood and Wood Products, 1965–95

Type	1965	1980	1995	Increase 1965-95
	(million cubic meters[1])			(percent)
Roundwood	2,231	2,920	3,331	49
Fuelwood and Charcoal	1,099	1,472	1,839	67
Industrial Roundwood	1,132	1,448	1,492	32
Sawnwood	384	451	427	11
Wood-based Panels	42	101	146	248
Pulpwood and Particles	238	370	419	76
Paper and Paperboard	98	170	282	189

[1]All units in million cubic meters except paper and paperboard, which are in
million tons.
Source: See endnote 43.

TABLE 3

Fuelwood and Industrial Roundwood Production and per Capita Consumption in Developed and Developing Countries, 1970 and 1994

	Share of World Population	Fuelwood and Charcoal			Industrial Roundwood		
		Production	Consumption		Production	Consumption	
	(percent)	(million cubic meters)	(cubic meters per 1,000 people)	(percent)	(million cubic meters)	(cubic meters per 1,000 people)	(percent)
1970							
Total	100	1,185	320	100	1,278	346	100
Developed	29	187	178	16	1,070	1,046	86
Developing	71	998	377	84	208	67	14
1994							
Total	100	1,891	336	100	1,467	262	100
Developed	23	191	156	11	1,051	864	73
Developing	77	1,700	386	89	417	92	27

Source: See endnote 44.

about 5 million cubic meters (enough to fill about 100,000 logging trucks) of wood every year just for curing or drying tobacco. For each kilogram of tobacco, almost 8 kilograms of wood is used in curing. The fuelwood collected by rural households is usually dead wood or trees grown for that purpose in mixed agricultural settings, and does not contribute to deforestation.[45]

The landscape of industrial timber production and trade has changed significantly during the past three decades. Both the volume and value of industrial production has increased. Since 1965 industrial roundwood production has risen by almost one third (See Table 2). This is the wood that is used to make lumber (sawnwood), paper, plywood, panels, and similar products. In the last 30 years there has been a major shift in emphasis from producing *quality* timber to producing *quantities* of fiber. Lower-quality products such as pulp for paper and wood-based panels like fiberboard and plywood have expanded far faster than traditional wood products like sawnwood.[46]

While the United States, Canada, and Russia have remained among the top five industrial wood producers in the world for over 40 years, China and Brazil only joined this group in the 1970s. These five countries produce over 57 percent of the world's industrial wood. Together, the top 10 (which also include Sweden, Finland, Malaysia, Germany, and Indonesia) account for over 71 percent of production. (See Figure 2.)[47]

Developed nations not only produce, they also consume a disproportionate share of the world's industrial roundwood. (See Table 3.) In fact, almost three quarters of the world's timber harvested for industrial use is consumed by the 23 percent of the world who live in developed nations, where populations are relatively stable. Their consumption per person is over nine times higher than that in developing nations, although the latter have increased their rate and share of consumption of forest products since 1970.[48]

Some countries have experienced dramatic reductions in wood production, for a variety of reasons. The Philippine

FIGURE 2

Top 10 Producers of Industrial Roundwood, 1995

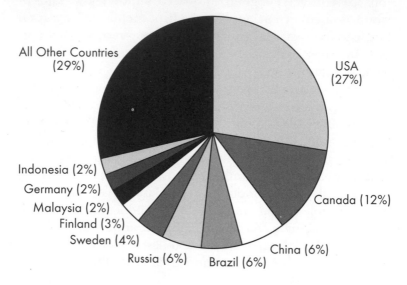

All Other Countries (29%)

USA (27%)

Indonesia (2%)
Germany (2%)
Malaysia (2%)
Finland (3%)
Sweden (4%)

Canada (12%)

Russia (6%)   Brazil (6%)   China (6%)

Note: Does not tally to 100 due to rounding.
Source: See endnote 47.

timber boom of the 1970s went bust by the 1990s when production fell by nearly three quarters and exports vanished—
along with 90 percent of its primary forest. Japan has shifted from timber production to timber imports and is now the
largest importer of raw logs, plywood, chips, and particles in
the world. In the former Soviet Union, legal and recorded
production dropped by nearly two thirds in the five years
between 1990 and 1995, due to economic, political, and
infrastructure difficulties during the breakup of the Soviet
Union. (For example, the government eliminated huge
transportation subsidies, without which it is not economically feasible to log remote interior regions).[49]

Trade in forest products—both legal and illegal—is a
strong economic force. Between 1970 and 1995 the value of
forest exports almost tripled in constant dollars. (See Table
4.) Although only 6 to 8 percent of timber and 25 percent of

pulp and paper produced is traded internationally, the legal and recorded trade of over $142 billion a year in timber, pulp, and paper makes forest products a valuable sector in the global marketplace. In recent years, tropical timber trade has received much attention because of rapid deforestation and the expansion of exports. (See Table 5.) But the temperate and boreal forests of developed nations still supply the majority of the world's industrial wood. In fact, they comprise 81 percent of exports by value.[50]

It is important to note that "trade" refers only to legal and recorded trade. While difficult to quantify on a global scale, evidence from numerous countries shows that there is a substantial volume of illegal trade that goes undocumented. For example, independent investigators estimated that one third of Ghana's logs are harvested illegally. In Brazil, the government calculates that 80 percent of the timber harvest in the Amazon is illegal.[51]

Many nations that have rapidly expanded timber production and trade have done so essentially by mining their forests—cutting at rates that far exceed the capacity of forests to regenerate. Between 1961 and 1995, Canada more than doubled timber production. During that time, Brazil and Malaysia expanded production of industrial wood over fivefold, Indonesia over sixfold, and they continue to cut at unsustainable rates. Indonesia and Malaysia have captured the lion's share of the expanding plywood trade. They now supply nearly two thirds of the world's exploding plywood and panel trade, up from less than 3 percent just 25 years ago. Indonesia and Malaysia are now the seventh and eighth leading forest exporters (by value) in the world. (See Table 4.) Not coincidentally, during the 1980s Indonesia, Brazil, and Malaysia together accounted for 53 percent of the world's forest loss.[52]

The demand for some forest products and in some regions has been growing very fast in recent decades. Wood panels—such as plywood and particle board—are relatively new products on the world market and their production has mushroomed. (See Table 2.) Malaysia and Indonesia moved

TABLE 4

Top 15 Forest Product Exporters, by Value, 1995

Country	1995 Exports	Share of World Total	Increase Since 1970
	($1,000,000)	(percent)	(percent)
World	142,344		185
Canada	27,787	19.5	157
USA	18,148	12.7	178
Finland	11,953	8.4	149
Sweden	11,582	8.1	90
Germany	7,800	5.5	468
France	5,837	4.1	374
Indonesia	4,728	3.3	1,267
Malaysia	4,232	3.0	256
Russia	4,028	2.8	22
Brazil	3,547	2.5	695
Austria	3,361	2.4	149
Netherlands	3,017	2.1	347
Italy	2,874	2.0	367
Belgium/Luxembourg	2,791	2.0	228
Norway	2,309	1.6	96

Source: See endnote 50.

aggressively into producing these panels for export in part because demand for forest products in Asia has been expanding far faster than anywhere else. Consumption of wood panels in Asia, for example, is rising at 5.5 percent per year, over three times the world average. Paper consumption in Asia has grown at 6.7 percent a year since 1980, over twice the world average.[53]

Global consumption of paper (including newspaper and paperboard) is increasing rapidly. From 1980 to 1994 it grew at 3.3 percent per year. The world uses more than three and a half times as much paper today as it did in 1961, and consumption is expected to grow by about half again by 2010. Currently, about 40 percent of the world's industrial

TABLE 5

Exports of Wood and Wood Products, 1970 and 1994

	Total 1970	Total 1994	Increase in Exports 1970–94	Share from Tropical Countries 1970	Share from Tropical Countries 1994
	(million cubic meters[1])			(percent)	
Industrial					
Roundwood	93.6	113.4	21	41	21
Sawnwood	57.4	107.6	88	9	10
Wood-based Panels	9.7	38.2	294	12	39
Plywood	4.8	17.7	269	15	71
Pulp[1]	16.9	31.6	87	<1	8
Paper and Paperboard[1]	23.4	72.7	211	<1	6

[1]All units in million cubic meters except pulp, paper, and paperboard, which are in million tons.
Source: See endnote 50.

wood harvest goes into making paper products; soon that portion is likely to be more than half. Fully 46 percent of the world's paper output is used to make packaging materials, while 41 percent goes to communications (newspaper, printing and writing papers) and 6 percent ends up in household and sanitary products.[54]

Paper consumption is not evenly distributed around the globe. (See Figure 3.) More than 66 percent of the world's paper output is used by the 16 percent of the population living in North America, Europe, and Japan. While global per capita use of paper stood at about 50 kilograms a year in 1995, the U.S. average was 341 kilograms per person (the world's highest), Japan's was 232, and Germany's was 200. In Brazil the figure was 35 kilograms, in China it was just over 29, and in India the average was less than 4 kilograms. Japan consumes nearly as much paper as China, which has almost 10 times as many people. Even though the rate of growth is now higher in developing nations, industrialized nations will continue to consume a disproportionate share in the future. (See Figure 4.)[55]

FIGURE 3

Paper Consumption and Population, Selected Countries and Regions, 1995

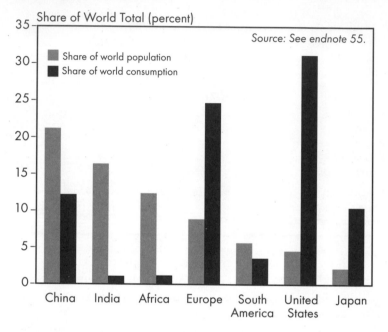

Share of World Total (percent)

Source: See endnote 55.

Share of world population
Share of world consumption

China India Africa Europe South United Japan America States

Wood accounts for about 58 percent of the fiber used to make the world's paper; only 6–7 percent comes from non-wood sources such as straw, sugar cane residue (bagasse), bamboo, kenaf, hemp, or cotton. Since 1970, the amount of recovered paper used in paper production has increased from less than one quarter to over one third today. (These portions can vary dramatically from country to country.) Although this has helped slow the growth in demand for wood pulp, increases in paper consumption have overwhelmed some of the gains made in recycling.[56]

Despite impressive strides in recycling, there is plenty of room for continued growth. Worldwide, over 41 percent of wastepaper is now recovered. In the United States, 45 percent of paper and paperboard is now retrieved, up from 29 percent in 1987, thanks in large measure to curbside recy-

FIGURE 4

Trends in per Capita Paper Consumption in Developed and Developing Countries, 1975–2010

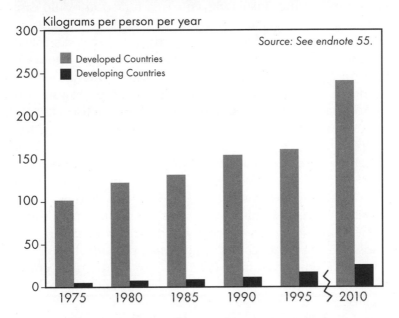

Kilograms per person per year

Source: See endnote 55.

■ Developed Countries
■ Developing Countries

cling programs launched by many local governments. The U.S. paper industry has a goal of 50 percent recovery by 2000, a standard already met in many countries. Germany currently recovers 67 percent of its wastepaper, South Korea 56 percent, and Japan 52 percent. The major obstacles to meeting the 50 percent target in the United States are low participation by offices and businesses, which are the largest source of high-quality wastepaper, and uneven enforcement of laws mandating recycled content in new products. Failing to recover and recycle paper also stresses waste disposal systems—in the United States, for example, paper products make up roughly 39 percent of the municipal solid waste stream.[57]

Lowering waste and excessive consumption by consumers would yield substantial benefits for forests and economies without sacrificing quality of life. Given the huge amount of paper consumed by the populations living in the United States, Western Europe, and Japan, to reduce their consumption and waste by even a small fraction would ease pressures on forests significantly. In the United Kingdom, for example, 130 million trees' worth of paper is discarded after use each year. In the United States, 14 million trees are used annually to make mail order catalogs. A German survey found that 98 percent of secondary product packaging was unnecessary, and a law designed to reduce such packaging waste was implemented. Nearly a fifth of all lumber in the United States goes into shipping crates and pallets, most of which are discarded after use. They account for 40 percent of all wood waste.[58]

Reducing the amount of wood burned as fuel is also possible. Since most of the live trees cut for fuel in developing countries are consumed by industrial and urban fuel users, shifting these sectors to methane, propane, and clean, renewable energy sources (such as wind and solar) could greatly reduce the pressures on the forests and improve air quality.[59]

Improving forest management techniques and reducing the excessive levels of waste during harvesting and processing are also important ways to ease pressures on the forests. There is ample scope for cutting waste and improving efficiencies at virtually every stage of production and consumption. One source of valuable wood is the high proportion of trees that are currently damaged and left on the ground in many commercial forest operations around the world—50 percent collateral damage is common. (Of course, when people simply torch forests to make way for plantations or agriculture, all of the wood may be wasted.) Better road placement and mapping of tree location and felling direction can reduce damage as well.[60]

Many species that are currently discarded or sold cheaply have high potential value. Better sorting and grading can help ensure that wood brings the best price and highest value use. In the tropics, only a few of the many hardwood species are currently marketed. One forest consultant stated that the "junk" woods used to make rough shipping crates for forest products in the tropics are often more valuable than the contents of the crate and have promise as valuable specialty woods. In the forests of the Pacific Northwest of North America, the yew tree once discarded as trash was found to yield taxol, an important cancer-fighting drug.[61]

Reducing the waste in processing also has enormous potential for diminishing pressures on forests and improving economic returns at the same time. In the United States, more than half the wood brought to a sawmill leaves as chips and sawdust, and about three fourths of this residue is used for pulp or fuel. In British Columbia, 40 percent of a log processed becomes lumber; the rest feeds the province's pulp mills. A study by IMAZON (the Amazonian Institute for Man and the Environment) in Brazil found that only one third of each harvested log is turned into sawnwood; the rest is discarded. Improved equipment maintenance and better training for workers could increase processing efficiency by 50 percent. If such advances were combined with better forest management practices, only one third as much forestland would be needed to produce the same amount of lumber. According to FAO, there has been some success globally in increasing product output with less roundwood input by recycling more materials and residues, and using more efficient technologies. There is room for even greater efficiencies. FAO suggests that if developing countries used this approach, consumption could grow "without placing unnecessary stress on the forest resource."[62]

Commercial exploitation of the world's forests is causing a number of fundamental changes and adding to the pressures already described. Intensive cutting as well as selective harvesting can result in simplification, fragmentation, and degradation of forests. So, too, does conversion to tree or agricultural plantations and pasture. These changes diminish the ability of forests to provide the full range of goods and services humankind depends on—from non-wood forest products to the regulation of water supplies and climate.[63]

Harvesting, roads, and conversion to other uses create a checkerboard of disconnected forest fragments. These fragments have more edge and less interior habitat—they are drier, sunnier, windier, more prone to wind and fire damage and invasions. And roads, highways, waterways, and pipelines all open the forest for exploitation and change brought by timber and agriculture operations, mining, hunters, landless settlers, and invasive species.[64]

The network of roads built into forests is extensive. In federally managed U.S. National Forests, for example, there are more than 600,000 kilometers of roads—enough to circle the globe nearly 15 times. One square kilometer of forest can contain up to 20 kilometers of roads. And in one large timber concession in Indonesia, 500 kilometers of logging roads themselves devoured 40,000 hectares of trees. In British Columbia, the length of forest roads increased by 42 percent between 1984 and 1993.[65]

Throughout Brazil, as elsewhere, the rapid and extensive deforestation of recent decades has been concentrated near roads. During the 1950s and 1960s the Brazilian government began building roads and infrastructure to spark migration and economic activity in its vast, untapped interior as part of a national integration program. The first big project was the highway from Belém to Brasília, the new national capital. Several million settlers were encouraged to relocate along the highway, and soon vast areas were cleared for cattle. At first, little of the wood was marketed, and billions of dollars' worth of timber was simply burned. As settlements, roads, services, and other infrastructure developed, however, and as transportation costs fell in the 1980s, timber extraction began to play a major role in the deforestation process. Since then, timber extraction in the Brazilian Amazon has increased 34 times. Transportation corridors are also facilitating the conversion of forests to produce agricultural commodities such as soybeans bound for Europe. As a result of the ambitious road-building and integration program, the area deforested in the Brazilian

Amazon expanded from 30,000 square kilometers in 1975 to at least 600,000 square kilometers today—more than 16 percent of the Brazilian Amazon—and twice as much area was affected biologically.[66]

Despite recognition by the Brazilian government and others that the earlier highways had serious and costly consequences, newer and bigger roads and infrastructure developments are being planned throughout the country, spurred by the booming global economy and easing of trade barriers. The goal is to increase trade within South America and with the rest of the world by constructing transportation and energy networks and expanding agriculture. These projects include highways, railroads, waterways, dams, power transmission lines, and gas pipelines. They would extend from the center of the Amazon north through Guyana and Venezuela to the Caribbean, west through Peru and Bolivia to the Pacific, and east to the Atlantic Ocean. Criticism of the projects has been levied by economists who question some of the supposed returns, and environmental and human rights advocates who are concerned about the destruction of forests, rivers, and wetlands and the violation of indigenous peoples' lands.[67]

In U.S. National Forests, there are [enough roads] to circle the globe nearly 15 times.

Over the last 20 years a new phenomenon is occurring in the moist tropical forests as a result of road building and logging: forest fires, previously rare in wet forest types, have become common. The fires that raged in Indonesia and Brazil in 1997 and 1998 are part of this new ecological pattern. On the other hand, in some forest types that are adapted to fires, such as some forests of North America, decades of fire suppression have allowed the buildup of tinder and thus increased the severity of major fires. In both cases, the balance between fires and forests has shifted as a result of human intervention.[68]

In Southeast Asia, the fires of late 1997 ignited regional

and global concern as smoke and haze darkened Indonesia, Malaysia, Singapore, Brunei, southern Thailand, and the Philippines for many months, shutting down schools, transportation, and businesses. Tens of millions of people sickened and hundreds died. Habitat for endangered species like the orangutan was devastated. By the Indonesian government's own admission, 80 percent of the fires were started by pulp and palm oil plantation owners to clear natural forest, by timber operations covering illegal activities. Many fires were also started by a government-sponsored project to clear and drain 1 million hectares of peat swamp forests for rice cultivation. The fires spread to at least 2 million hectares of forest and underground peat deposits and reignited in 1998. A drought induced by El Niño aggravated the situation. Enormous amounts of carbon dioxide—perhaps as much as emitted by all of Western Europe in one year—were added to the atmosphere. Losses from the 1997 fire were projected by WWF to exceed $20 billion when lost productivity, tourism, timber, health impacts, and other costs are tallied.[69]

A major force behind the large-scale forest exploitation and infrastructure developments just described are major logging corporations, which have long been heavily involved in the timber trade, and which are now expanding their reach. Although most internationally traded timber comes from temperate and boreal forests, and is harvested by companies from those nations, firms based in southern countries, especially in Asia, play a growing role. Japan, South Korea, China, and Taiwan are major timber importers and processors, and Malaysia and Indonesia have become important global exporters.[70]

As some Asian nations have depleted their forest resources, they have turned elsewhere to satisfy their domestic consumption needs and the demands of their forest industries. Companies that have grown wealthy from domestic timber exploitation are using their capital to expand elsewhere. Some of the operations are in northern temperate and boreal forests—such as the logging of Siberian forests by South Korean firms and of Canadian forests by Japanese companies—but much of the expansion occurs in southern nations. One Malaysian company controls 60–85 percent of the logging concessions in Papua New Guinea (PNG). In 1996 alone, the area of Amazon forest under concession to Asian timber companies quadrupled to more than 12 million hectares. Malaysian, Korean, Indonesian, and Chinese companies also operate in Suriname, Guyana, Belize, PNG, Solomon Islands, Africa, Cambodia, and Myanmar, among others.[71]

There are several reasons for the rising influence of roving international companies. First, in the past decade international trade restrictions and tariffs have been eased and global and regional trade agreements have expanded. Domestic policy measures—such as subsidies to mining and timber-processing industries, and give-away concessions and monopolies—have helped create powerful firms with the capital and connections to look beyond their home countries for raw materials and higher profits. By operating in nations with less restrictive laws, lower fees, and lax enforcement, timber companies can reap high profits from their legal—and sometimes illegal—harvest.[72]

The power of timber companies and the desperate economic situation of countries allows them to dictate terms.

The size and power of the timber companies and the often desperate economic situation of host countries or communities allows the companies to dictate favorable terms. In the Solomon Islands, for example, landowners were paid $2.70 per cubic meter for timber that foreign companies then sold for $350 per cubic meter. In Suriname, companies from Indonesia, Malaysia, and China proposed investments of more than $500 million—an amount nearly the size of that nation's annual economic output but just a small fraction of the timber's value. While substantial international public pressure has delayed release of the concessions, significant amounts of illegal logging have proceeded in those areas, financed by the foreign companies, according to knowledgeable sources. In Belize, a Malaysian timber company was granted a concession for over 89,000 hectares at about \$1.34 per hectare. What may appear to be a short-term boost to the domestic economy (and to the few individuals who benefit legally or illegally) is far outweighed by the economic and ecological losses that last long after the logging operations have departed.[73]

What Governments Do

Domestic laws, policies, and attitudes have enormous influence on how forests are managed, on who benefits from their use and who suffers from their misuse. Where governments control a significant portion of the forest estate—as in Canada, where 94 percent of forestland is publicly owned, or in Indonesia, where the state controls 74 percent—the role of government is obvious. But even without direct ownership, government trade and economic policies, management regulations, and agriculture and land tenure policies—and the ability to control corruption and cronyism—exert significant influence over the fate of forests.[74]

One miscalculation that often leads to over-exploitation of this life-support system is the undervaluing of benefits provided by intact natural forests. Forests are often viewed as vast, uninhabited spaces that are valuable only when converted to agriculture or mined for timber. This mindset leads to overestimating the economic benefits of forest exploitation or conversion. Still, governments usually underprice timber and other forest products. The combined effect is to encourage rapid forest exploitation, depletion, and waste, and to sacrifice public revenues and benefits from intact forest.[75]

Forests are routinely sold at prices far below what the timber alone is worth. In Canada, stumpage rates—the fees paid for each cubic meter harvested from government lands—are half of what they are in the United States, with large companies paying even less than small ones. And in Indonesia, an independent assessment of timber concessions by economists concluded that in 1990 alone the government collected less than one fifth of the potential revenues—a loss of $2.5 billion.[76]

Just as a small landowner will sell a few trees for cash during hard times, governments often look to their forests as a standing asset that can be liquidated to solve financial problems. In Russia, some cash-strapped municipalities are paying creditors with forestland, and its far east has been opened up to resource exploitation by outside companies. The economically desperate South American nations of Suriname and Guyana have been considering bids that would give away half of their forests to Asian timber companies for pennies per hectare. When Indonesia's military government came to power in the late 1960s, it took over a country with massive debt and high inflation. The new leaders put in place a series of policies—from underpricing logs, to subsidizing timber processing, to give-away concessions—that rewarded political and military supporters—but precipitated the rapid depletion of Indonesia's forests. By 1991, concessions to 41 percent of the nation's forestland had been granted to a small number of companies.[77]

Underpricing and lost revenue from timber on public land even in wealthy countries can be so substantial that governments in effect pay private interests to take public timber. In the United States in 1995, for example, 117 of 122 National Forests returned less money to the treasury than the Forest Service spent preparing the concessions for sale. From 1992 to 1994, the timber sales program lost $1 billion in direct costs alone (such as road-building and mapping), not including the costs of reforestation, stream erosion, and lost fisheries and water supply, recreation, and so on. The most heavily subsidized logging is in the coastal rain forests of Alaska. Even though timber sales from federal lands have turned a profit in only 3 of the last 100 years, Congress repeatedly mandates high harvest levels.[78]

Governments also underprice their forests by levying a flat charge for timber rather than differentiating between more and less valuable species. And they may base fees on the volume of timber removed from a site rather than the volume available. This policy encourages concessionaires to take and pay for only the most valuable species. Meanwhile, more forest is degraded and less revenue is returned to the government. Short concession terms, where the loggers have no incentive to ensure that forests regenerate because they will not be there to re-harvest, also encourage a cut-and-run approach.

Governments often promote timber extraction to create jobs. Ironically, many of these extractive industries generate relatively little employment, especially when compared with other options for forest use. For example, the U.S. National Forests are currently managed primarily for timber supply, despite the fact that logging adds only 76,000 jobs and $3.5 billion to the GDP (gross domestic product) while recreational use of the national forests generates nearly 2.6 million jobs and adds $97.8 billion to the GDP. Shifting from logging to recreation would also remove from the market the cheap timber that drives down the prices private landowners can get for their timber.[79]

One way that governments have attempted to raise revenues and promote employment from wood product industries has been to encourage domestic timber processing in order to add value to logs. This can also be a way to reduce the pressure on forests by extracting more value and jobs from less wood. British Columbia's Vernon forest district found sustainable forestry practices generated three times as many jobs in the forest as clear-cutting, and that individually grading and sorting the logs added more jobs and allowed the logs to be sold for an average of four times the standard prices. Because the wood was available to small buyers as well as large companies, value-added processing industries—from log home builders to fine furniture and instrument makers—now had a source of wood. In the southeast United States, a recent study found that for every $1 million invested in pulp mills only one new job was generated, while the same amount invested in furniture manufacturing created 40 jobs.[80]

Unfortunately, in some cases, encouraging a rapid shift to domestic processing has actually backfired, reducing revenues and encouraging deforestation. In Indonesia, for instance, the government banned the export of raw logs in 1985 and gave heavy financial incentives to stimulate the development of processors such as plywood mills. Without these inducements and tax concessions, timber processing in Indonesia would not have been profitable. In just a decade, they expanded from producing almost no plywood to manufacturing $3.7 billion worth, 47 percent of the world's plywood trade in 1994. As successful as this strategy may seem at first glance, the effort to add value to timber exports backfired as logs were actually reduced in value in inefficient mills, a plywood cartel was established, and demand for wood soared to meet the mills' demands. Even with illegal logging, some mills cannot operate at full capacity. Despite clear timber shortages and a 1993 World Bank assessment that found harvests 50 percent above sustainable levels, the Indonesian government continues to encourage domestic processing, and plans to raise harvest levels by 57 percent.[81]

Another manifestation of the failure to recognize the value of intact forest is laws that grant ownership and tax and credit benefits and subsidies to those who "improve" forest by clearing it. A series of policies begun in the 1960s to spur investment in the interior of Brazil sparked the deforestation that has affected so much of that nation's land. Roads built deep into the country's interior, generous tax holidays, credit with negative interest rates, and other subsidies encouraged the conversion of millions of hectares of forest to cattle ranches that would otherwise not have been profitable. By 1980, at least 72 percent of the conversion detected by satellite was for cattle pasture. After 1990, subsidized ranches caused four times as much deforestation as nonsubsidized ranches did, even though a quarter of the pasture was already abandoned because the forest soil could not support that land use. Brazil lost more than valuable forest. By 1988, the fiscal cost to the country of all 470 subsidized ranches was $2.5 billion. Despite some tax reforms, taxes are still higher and less credit is available on land with forest cover, and Brazil is pushing even more ambitious infrastructure and agriculture expansion plans.[82]

Governments also use forests as safety valves, to reduce pressure in heavily populated places by siphoning people off to new areas. Indonesia's transmigration program moved settlers from Java to the nation's less populated islands. During the 1970s and 1980s, about 6 million people were relocated. Nearly all these people were settled in forested land, much of it already occupied by native tribes like the Dayak. An estimated 3 million hectares—5 percent of the country's forest—were converted to agriculture during this scheme. The cost to the government was about $10,000 per family, an enormous amount in a nation where the per capita gross national product was only $530. Despite the massive infusion of funds, many of the settlements have already been abandoned, and the people have moved on or returned to Java. Similar resettlement programs have also failed in Malaysia and Brazil.[83]

The Indonesian government's transmigration settlements also played a role in the 1997 fires. A government scheme to drain and clear 1 million hectares of peat swamp forests in southeastern Kalimantan to convert it to wet rice cultivation and bring in 300,000 families led to the area's burning. Not only is peat swamp unsuitable for such cultivation, draining the swamps made them highly flammable.[84]

Because forests are often seen as vast uninhabited spaces, forest dwellers—if acknowledged at all—are usually considered encroachers who impede development. Rarely is the distinction made between *shifting* cultivators—who have a long history of successful forest management, like the Dayak of Borneo—and *shifted* cultivators, settlers who have been relocated to forest areas often without knowledge of how the forests should be managed.[85]

Few forest communities have been successful in gaining recognition for their customary rights to the very resource they have often managed sustainably for generations. Their occupancy has been banned in some cases, and disregarded in others. Even when laws are passed allowing for the demarcation of tribal lands (as in Brazil) or community forest management (as in India), they are often not enforced, and encroachment by outsiders is tacitly allowed. In Brazil and Venezuela, hard-won indigenous reserves have been invaded by miners, loggers, and settlers. Sometimes they will rush to stake a claim on land in anticipation of indigenous claims. Brazil has recently cut funds for demarcation. In Guyana, a decades-old law states that logging and mining concessions cannot overlap with officially recognized Amerindian lands—which cover 16 percent of the country. However, as many as 41 Amerindian communities have not yet been officially recognized or had their lands titled, placing them in a precarious position as new timber and mining concessions are granted. Only since 1994 has the Canadian government begun to take steps to initiate treaty negotiations and recognize the rights of First Nations to manage the lands they live on. In nation after nation, communities' loss of their ability to control access to their forestlands has hurt both people and nature.[86]

Communities' loss of control of their forests has hurt both people and nature.

In Indonesia, the government created a real-life "tragedy of the commons" when it declared in 1967 that it had sole legal jurisdiction over the nation's forests—74 percent of the land area. Customary rights to common areas, which had evolved as a complex and sustainable management system over many generations, were not legally recognized. The government, which claimed the authority, was unable to police the nation's vast forests, and the communities who are in the forest no longer had the power to stop exploitation by outsiders. One analysis concluded that "the traditional...rights of millions of people...have been handed over to a relatively small number of commercial firms and state enterprises."[87]

Little of the economic gains from forest exploitation return to the communities who have lost access to forest resources. In fact, their standard of living has declined. Most of the profits benefit a few powerful industries or families (many of whom helped bring President Suharto to power and were granted generous concessions as reward). Similarly, the liquidation of 90 percent of the Philippines' primary forest during the Marcos regime made a few hundred families $42 billion richer, but impoverished 18 million forest dwellers.[88]

Domestic policies can also have unintended consequences for the forests of other nations. After the devastating floods and landslides that originated in its deforested highlands in 1985, Thailand enacted a logging ban. Although legal domestic logging ended, domestic consumption did not, spurring logging (much of it illegal) in neighboring Myanmar and Cambodia. Some of the activity was aided by the army. Indonesia's and Malaysia's policies that encouraged rapid and wasteful exploitation of domestic timber also fed the growth of large companies and overcapacity in the industry. Now the companies travel the world looking for timber to feed their mills and coffers.[89]

Even when there are good forest laws and policies on the books, all too often, governments lack the capacity or the will to enforce them. Logging beyond the boundaries of concessions and in sensitive river and stream areas, tax evasion, and falsification of boundaries, log volume, and grades are all common practices in timber concessions around the world. So too are the harvesting of protected species, exceeding quotas, and not mapping and reforesting as required. Penalties, such as they are, are usually too light and too rarely applied to work as a genuine deterrent. Corruption can be found from forest agents to top officials. Companies see bribes and fines as a very minor business cost.[90]

While the exact amount of illegal logging and trade is difficult to determine, numerous examples indicate that it is significant and widespread. The Brazilian government reports that 80 percent of timber extraction in the Amazon is done illegally. In Russia, logging may be occurring illegally on as much as 12 million hectares a year compared to the 2 million hectares of legal logging in official estimates.[91]

Many nations lose significant portions of their forests and potential revenues as a result of failure to enforce their existing laws. Papua New Guinea's losses from unmonitored log exports alone, for example, were estimated at $241 million a year in 1994. According to investigations by Friends of the Earth International and in-country partner organizations, Cameroon lost over 50 percent of its potential tax revenues during this decade because most of the large timber companies participate in illegal trade. In Ghana, about one third of timber is harvested illegally. If this situation continues, Ghana will lose $65 million a year and 10,000 jobs, and the 12 most popular commercial timber species will become extinct within a decade. The situation is especially critical because 95 percent of the nation's forests outside of reserves have already fallen.[92]

In Cambodia, the amount lost to the national treasury as a result of illegal logging alone is equal to the entire national budget. According to the environmental and human rights investigative group Global Witness, Cambodia's prime ministers and the military control the nation's forests and timber trade—most of which is illegal. Profits bypass the official budget and go directly to a parallel budget that funds the factions in the ongoing civil war. Timber concessions for the nation's remaining forests were awarded in 1995, also in violation of the law. The Khmer Rouge guerillas have also been illegally logging areas under their control. Based on the amount of timber known to have been exported in 1995 and 1996, for example, $400 million should have been generated, yet only $10 million came to the treasury. The losses to the people of Cambodia who depend on the forests and fisheries are far higher. The forests are expected to be depleted in the next decade, and the Tonle Sap—the great lake, which is one of the world's richest fishing grounds and the source of much of the nation's water and protein—will be silted up in 25 years if deforestation continues.[93]

Nations with weak laws or enforcement capabilities or that are prone to corruption make vulnerable targets for domestic or foreign companies looking for cheap timber. The Brazilian Environmental Agency (IBAMA) has just 650 technicians, 120 vehicles, and 30 boats to inspect the entire Brazilian Amazon, over 3.8 million square kilometers—an area one and a half times the size of Western Europe. Even if it had the resources, IBAMA has had no statutory authority to enforce forest laws, collect fines, or seize illegally harvested timber since 1989. When the law reinstating this authority was finally passed in early 1998, many of the toughest provisions had been eliminated under pressure from ranching and timber interests. Further, the budget for forest protection was reduced by 64 percent and permits are being granted for logging in national forest reserves. Suriname— where the forest service has few staff, and just one vehicle to monitor nearly 150,000 square kilometers of forest—has little capacity to enforce even minimal contractual and environmental standards on the timber concessions proposed for up to 40 percent of the country.[94]

Even in nations with a relatively well-staffed, -funded, and -monitored forest service, enforcement problems can occur. In the United States in the early 1990s, it was discovered that timber companies were stealing hundreds of millions of dollars' worth of trees from federal lands each year, sometimes with the knowledge of Forest Service employees. The Forest Service eventually won a multimillion-dollar lawsuit in court, but recovered only a small fraction of the value of the lost timber.[95]

Government policies and enforcement can be easily influenced or subverted by powerful interests. Strong ties between politicians and their families, the military, and extractive industries thrive in many nations today, including Malaysia and Indonesia. Companies with ties to

Indonesia's president were removed from the government list of those implicated in the devastating fires of 1997–98. Money from Indonesia's reforestation fund is routinely diverted by President Suharto for projects that benefit loggers and even nonforest uses, such as aircraft manufacturing. In 1997, he ordered $115 million transferred from the fund to build a paper factory for timber magnate "Bob" Hasan, a business partner and friend of the presidential family, who also had a hand in crafting Indonesia's forest policy. During the 1997–98 currency crisis and International Monetary Fund (IMF) bailout, the IMF's audit of the reforestation fund found no money left to fight the devastating fires—it had been siphoned off to prop up the "national car" company of the president's son.[96]

In Canada, the timber industry is a powerful force in the economic and policy arena. More than $32 billion worth of forest products are exported each year—making Canada the largest forest exporter by value in the world. Government-owned "crown lands" account for 96 percent of the forests, and more than three quarters of all timber revenues from Canadian crown lands come from the province of British Columbia. Forests are leased to timber companies, and high-volume logging is stipulated by law in the contracts. Just 10 companies hold long-term leases to over half the province's rich timber resource.[97]

Liquidation of all old-growth primary forest has been explicit government policy in British Columbia since 1945. Cutting has more than doubled in the last 30 years. Even by the government's own estimates, the total cut currently mandated by law (the annual allowable cut) exceeds long-term harvest levels by more than 20 percent. Government investigations "made it clear that provincial cut levels were being driven by forest industry demand for timber rather than a clear understanding of the productivity of the forest land base," according to an independent study by BC Wild.[98]

Widespread international concern over the rapid degradation of British Columbia's rich temperate rain forest through this industrial clear-cutting earned the province the

label "Brazil of the North." Two thirds of Canada's coastal
rain forest, a rare and threatened ecosystem, has already
been degraded by logging and development. British
Columbia also serves as an important habitat for salmon—
of which 140 stocks are already extinct and 624 are at high
risk. Salmon depend on intact forested watersheds and
streams for survival and reproduction.[99]

In the 1980s and 1990s, Clayoquot Sound on Vancouver Island attracted intense local, national, and international concern because of its size, unique qualities, and the
rapid clear-cutting that was taking place—threatening its
ecology and economy. In 1993 the provincial government
established an independent Scientific Panel for Sustainable
Forest Practices, made up of scientists, foresters, local people, and First Nations (the native Indians) to study how
forestry was being conducted and to make recommendations on how to "develop world-class sustainable forestry
practices for Clayoquot Sound."[100]

The panel asserted that "the key to sustainable forest
practices lies in maintaining ecosystem function." Their
1995 report to the government made 120 recommendations
that—if followed—would dramatically change existing forest practices. These recommendations included basing planning on watershed conditions and capabilities rather than
pre-set volumes, designating critical forest reserves before
selecting harvest areas, focusing planning on the ecosystem
elements and processes that should be maintained rather
than on the resources to be extracted, incorporating scientific and traditional ecological knowledge, and making management and regulations adaptive to changing circum-
stances. Important recommendations included recognizing
First Nations' values, their knowledge, and their stake in the
region, and implementing co-management of the forests
with them. The panel emphasized that changes were needed
in "both the philosophy of forest planning and management
and the way that forest practice standards are created."[101]

The government pledged to adopt and implement
these ground-breaking recommendations. While in theory

they applied only to Clayoquot Sound, the problems identified mirrored those in the rest of the vast province and the proposed solutions were widely applicable. In 1995, British Columbia enacted a Forest Practices Code that was supposed to be the linchpin of the government's efforts to create "world-class" sustainable forestry that would "stand up to world scrutiny."[102]

Unfortunately, after the Forest Practices Code became law, a 1997 audit by Canada's Sierra Legal Defense Fund of timber-cutting plans for 10,000 forest blocks approved by the Ministry of Forests found a vast difference between the letter of the code and the plans approved. Contrary to the code, clear-cutting was the harvest method on 92 percent of the blocks, including landslide-prone slopes; 83 percent of streams were clear-cut to the banks; fish-bearing streams were misclassified or unidentified by the companies; and destructive yarding—dragging logs through streambeds—was common. The annual cut was not reduced as promised, and harvest blocks were more than twice the allowable size. None of the special areas for wildlife and biodiversity protection or old-growth management called for in the code had been designated. Instead of the million-dollar fines promised, only 9 of 120 levied were over $10,000.[103]

Liquidation of old-growth forest has long been government policy in British Columbia.

These findings and others led many foresters, as well as environmental and First Nations groups, to conclude that the Forest Practices Code's standards and enforcement were inadequate to ensure long-term sustainable yield—much less ecosystem health—and that too much of the responsibility for identifying and protecting sensitive areas was left to the discretion of logging companies, who neglected their obligation. Despite the lax rules and apparently laxer enforcement, the industry complained that the code was too burdensome and was hurting its profits and market share. In June 1997, the government eased the Forest Practices Code.

Moreover, efforts to gain leases to practice ecoforestry have been almost entirely rebuffed as the government continues to assign leases to the large industrial timber companies.[104]

Sustainable Forest Management

Managing forests exclusively for timber commodities and converting the land to other uses has reduced or curtailed the ability of forests to provide many other products and services that people depend on. Still, it is clear that the world will continue to need timber products, and that much of that need will be satisfied through commercial forest management. Thus foresters, ecologists, and economists are seeking to transform forest practices into sustainable management. Unfortunately, when many foresters use the term "sustainable forestry" today they usually mean "sustained yield"—that is, a continuous supply of timber and fiber. Even by that weak standard, forestry has been failing so far to sustain its resource base. In Canada, the world's second largest wood producer, British Columbia claims to have world class sustainable forestry, yet cutting levels exceed sustainable yield by more than 20 percent (in some areas cutting rates are double the sustainable yields). In the late 1980s, when the last estimate was made for the tropics, less than one tenth of 1 percent of tropical forests were managed for sustained yield. In the last few years, some in the industry have accepted principles of sustainable forestry that incorporate other goals, yet for most, maximizing timber production remains the bottom line.[105]

Under the principles of sustainable forest management (SFM), on the other hand, forests are managed as complete ecosystems to supply a wide array of goods and services for current and future generations. As Kathryn Kohm and Jerry Franklin of the University of Washington College of Forest Resources put it: "If 20th century forestry was about simplifying systems, producing wood, and managing at the stand level, 21st century forestry will be defined by understanding and managing complexity, providing a wide range of ecological goods and services, and managing across broad landscapes...managing for wholeness rather than for the efficiency of individual components." In recent years, progress has been made in understanding the complexity of forests, defining SFM, and describing how it can be applied in various forest types and nations. Along with the foresters, ecologists, and economists, several intergovernmental bodies have been developing international criteria and indicators to assess conditions in tropical, temperate, boreal, and dry forests, such as the Helsinki and Montreal Criteria and Indicators of Sustainable Forest Management, the Tarapoto Proposal of the Amazonian Cooperation Treaty, and the Dry-Zone Africa Initiative.[106]

While the concept of sustainable forest management continues to evolve, some elements are common to most definitions. First, forests should be managed in ways that meet the social, economic, and ecological needs of current and future generations. These needs include non-timber goods and ecological services. Management should maintain and enhance forest quality, and look beyond the stand to encompass the much larger landscape so that biodiversity and ecological processes are maintained. When trees are cut, the rotation period should follow the longer, natural cycle of a forest rather than a shorter, financial cycle.[107]

A sustainably managed forest would mirror the conditions in natural forests that are heterogeneous, with many species, ages, and sizes. Natural disturbances are enabled and mimicked. Sensitive areas like streams and important habitat such as dead tree snags are protected. Since forest species are interdependent, species such as fungi and insects that were once considered pests are kept because they are important to ecosystem functioning. Finally, sustaining forests requires the active and meaningful participation of all stakeholders, especially local communities.[108]

That forests can be used by people in ways that meet the principles of SFM has been demonstrated in forests around the world, from those managed by indigenous peoples and small landowners to those run by large timber companies. The "rattan gardens" of Indonesia and Malaysia have provided employment and income for hundreds of thousands of forest dwellers who manage the forest for rattan harvests in ways that are virtually invisible to the outside observer. (Many of these rattan gardens in Indonesia were destroyed in the 1997–98 forest fires.)[109]

The Menominee Tribe of Wisconsin has been practicing sustainable industrial forestry for many decades on about 90,000 hectares, and became the first source of certified sustainable timber in North America. Their practices sustain a high-quality and high-volume forest and produce a higher-quality wood than is generally available in the market, which allows them to charge a premium price. They are also sustaining forest health and the well-being of communities.[110]

Sustainable forest management is possible even in larger industrial operations. In Sweden, AssiDomän, one of Europe's largest forest products companies and one of the world's largest forest owners, has already had half of its 3.2 million hectares of productive forest certified as well managed by the Forest Stewardship Council, and expects to have the rest certified in 1998.[111]

A number of studies have shown that commercial SFM is also possible in the tropics. In the mangrove and peat swamp forests of Malaysia, for example, low-impact sustainable forest management with only minor timber harvests, which allowed other uses and benefits to continue—such as water supply for agriculture and domestic uses, carbon storage, non-timber products like fish and rattan, etc.—produced at least the same amount of financial returns per hectare and in many cases greater returns than just cutting all the wood. The difference was that in unsustainable uses, the beneficiaries were a small number of timber concession holders, and in the sustainable options, benefits were more widely distributed and longer lasting.[112]

Practicing sustainable forestry can incur some additional costs for industrial forest owners and managers at the beginning. The Sustainable Forestry Working Group estimated that initial costs may be 10–15 percent higher than standard management practices but diminish over time. However, they figured that much of these costs are offset by savings in efficiency. Where timber concessions are operating on public lands, practicing SFM may mean that concessionaires can harvest less timber initially. But these "losses" to concessionaires may be counterbalanced by higher revenue over a longer time period and by gains in public benefits such as improved water supplies, and so forth.[113]

At the same time that foresters and ecologists have been redefining the science of forestry, many consumers have indicated they want their buying habits to be part of the solution to forest decline rather than its cause. This concern is shared by a growing number of commercial buyers and retailers. In response, "ecolabels" for forest products and self-certification schemes by industry and government have proliferated that create confusion in the marketplace, and some amount to little more than "greenwashing." Many claims have been made—"five trees planted for each one harvested," "made from plantation grown trees," "environmentally friendly," "sustainable," for example. Unsupported claims also put producers using more sustainable methods at a competitive disadvantage.[114]

For claims to be meaningful and credible, independent auditing and verification are necessary. To accomplish this, environmental groups, foresters, timber producers and traders, indigenous peoples' groups, and certification institutions established the Forest Stewardship Council (FSC) in 1993. This group has developed "Principles and Criteria for Forest Stewardship" (see Table 6) that apply to tropical, temperate, and boreal forests managed for forest products. Detailed standards based on these principles are being developed by national and local councils. FSC accredits certifiers who, at the request of companies wishing to use the FSC logo, audit forest management practices and certify products for the entire chain of custody, from forest to processing. By

TABLE 6

Principles and Criteria for Forest Stewardship

- Forest management shall respect all applicable laws of the country in which they occur, and international treaties and agreements to which the country is a signatory and comply with all FSC Principles and Criteria.

- Long-term tenure and use rights to the land and forest resources shall be clearly defined, documented, and legally established.

- The legal and customary rights of indigenous peoples to own, use and manage their lands, territories, and resources shall be recognized and respected.

- Forest management operations shall maintain or enhance the long-term social and economic well-being of forest workers and local communities.

- Forest management operations shall encourage the efficient use of the forest's multiple products and services to ensure economic viability and a wide range of environmental and social benefits.

- Forest management shall conserve biological diversity and its associated values, water resources, soils, and unique and fragile ecosystems and landscapes, and, by so doing, maintain the ecological functions and the integrity of the forest.

- A management plan—appropriate to the scale and intensity of the operations—shall be written, implemented, and kept up to date. The long-term objectives of management, and the means of achieving them, shall be clearly stated.

- Monitoring shall be conducted—appropriate to the scale and intensity of forest management—to assess the condition of the forest, yields of forest products, chain of custody, management activities, and their social and environmental impacts.

- Primary forests, well-developed secondary forests and sites of major environmental, social, or cultural significance shall be conserved. Such areas shall not be replaced by tree plantations or other land uses.

- Plantations shall be planned and managed in accordance with these Principles and Criteria. While plantations can provide an array of social and economic benefits, and can contribute to satisfying the world's needs for forest products, they should complement the management of, reduce pressures on, and promote the restoration and conservation of natural forests.

Source: See endnote 115.

using globally consistent principles and an easily recognizable single label, FSC certification can help ensure consumer confidence and improve market access for timber from well-managed forests around the world. Companies as varied as the Menominee Tribal Industries and Collins Pine in the United States, AssiDomän in Sweden, communal forests in Mexico, and industrial plantations have been certified.[115]

A promising initiative, the FSC has had a small but growing impact in its first few years. The area certified has expanded to over 6.3 million hectares in 20 nations in early 1998. The amount of certified wood traded doubled from 1994 to 1996. Since worldwide demand for certified wood exceeds supply, there is room for considerable expansion.[116]

Companies that pledge to produce, market, and purchase wood products certified to FSC standards have said they do so because they believe their customers expect it and because it makes good business sense. Commitment by industry can in turn promote better forest management by their suppliers. For example, the 75 companies in the "UK-1995 Plus" buyers group, which have pledged to phase out wood products that do not come from well-managed forests as defined by FSC principles, represent about 25 percent of the U.K. market. B&Q, the largest do-it-yourself store in the United Kingdom, and Home Depot in the United States are committed to being outlets for certified wood. With a consistent market for their products, producers can take the step of adopting SFM.[117]

So far, the greatest impact of certification has been in the United States and Europe—which is significant because these regions are major producers as well as consumers. In the important Asian markets where consumption is growing at rates several times faster than the world average, certified forest products have barely surfaced. The certification concept has only just been introduced in Japan, which is a major importer and consumer—over one third of all the raw logs traded internationally end up there, as do one fifth of the plywood and nearly three quarters of the chips used to make paper. Raising awareness and demand for certification in this region—combined with reducing overall consumption—could have a major positive impact on the world's forests.[118]

Certification is not a panacea, of course. It cannot substitute for reducing wasteful consumption or for sound legislation and policies. It does, however, provide a voluntary market-based approach to fostering sustainable forest management and trade. It also provides a positive alternative to bans, which can boomerang and make alternative land uses, such as ranching or agriculture, more profitable than maintaining forests. These voluntary standards can complement the other national and international initiatives noted earlier. Challenges for the future include making certification workable and affordable for small forest owners (who predominate in Europe, for example) and labeling products by species and country of origin.[119]

Cultivating a New Relationship

Governments, businesses, citizens, and nature pay a high price for the continued mismanagement and undervaluation of forests. With the demand for forest products expanding and forests declining in area and quality, cultivating a new relationship with forests—one that ensures conservation, sustainable use, and the fair and equitable sharing of their benefits—offers a way out of the impasse.

The new relationship will necessitate some changes in prevailing attitudes and practices, but we have already demonstrated that these are possible and indeed they have already begun. The elements of the new relationship include halting forest degradation and conversion, improving research and monitoring, restoring forest health, improving management, reducing waste and overconsumption combined with making consumption more equitable, getting the market signals right, ensuring community participation in forest management, and reforming and strengthening national policies as well as international agreements.

Furthermore, restoring or establishing the rule of law and eliminating the widespread corruption and cronyism that stand in the way of achieving these goals are essential.

An overarching goal of the new relationship is to halt degradation of remaining primary forest and restore forest cover and health. Timber mining and clearing natural forests to establish tree plantations or agricultural land has no place in the twenty-first century. Sustainable forest management—a long-standing practice in many communities, and a more recent one in some commercial forest enterprises—needs to be expanded in scale. A proposal to raise the area under certifiable sustainable management from over 6.3 million hectares in February 1998 to 200 million hectares by 2005 (about 6 percent of today's forest area) has recently been endorsed by environmental and business groups as well as by the World Bank.[120]

There is still much to learn about forest species, functioning, and dynamics and about the best management practices. Funding for forest-related research is a small fraction of that for agriculture research, and both are inadequate to meet the challenges of tomorrow. Many nations do not have the budgets or resources to monitor or manage their forest estates adequately. More investment and a building up of these nations' capacities for forest management would reap substantial benefits by ensuring the long-term health of the world's forests.[121]

Improving the monitoring of global forest conditions and threats provides a key opportunity for international cooperation. As noted earlier, major weaknesses exist in the data on forest area and conditions gathered by national governments and FAO. In order to assess the state of the world's forests more accurately, existing monitoring mechanisms need to be strengthened, data collection procedures and classifications need to be improved, satellite monitoring used, in-country capacity strengthened, and an independent monitoring mechanism such as the Global Forest Watch put in place.[122]

One strategy for maintaining and restoring healthy

forests is to expand the protected areas network to ensure adequate ecological representation of all forest types. Protected areas today serve a much broader array of social and ecological functions than the parks of the past, which were seen simply as places of beauty. The World Wide Fund for Nature and the World Conservation Union have proposed that a minimum of 10 percent of each forest type be in protected areas by 2000. Currently only 6 percent of the world's forests have some protected status, and in many cases that protection exists in name only.[123]

Rehabilitating and restoring forests will become increasingly important as nations seek to regain the social and environmental benefits that forests provide. To be successful, they will need to shift from current practices of planting large areas of single (often exotic) species with little consideration to local needs or environmental services to using instead a mix of native species that provide multiple benefits. Preventing the accidental or intentional introduction of exotic species, which can wreak havoc on ecosystems, is also an important part of restoring forest health. Intensive plantations have a role to play in meeting the demand for forest products, if they follow these guidelines and are established on already degraded land. Of course, one of the potential benefits of plantations—reducing pressure on natural forests—does not come about if they convert natural forest or push people who depended on the land further into remaining forestland.

There is enormous potential for reducing waste and improving efficiencies at virtually every stage of production and consumption—improvements that diminish pressures on forests and improve economic returns at the same time. For instance, many species that are currently discarded or sold cheaply have high potential value. Globally, there has been some success in increasing product output with less roundwood input by recycling more materials and residues, and using more efficient technologies. There is room for even greater efficiencies. Post-consumer recycling has been expanding and there is ample room for continued growth.[124]

Lowering waste and excessive consumption by consumers would yield substantial benefits for forests and economies without sacrificing quality of life. Unless industrial nations reduce waste and overconsumption as developing nations expand their use of processed products, even greater pressures will be placed on the world's forests. If, for example, everyone in the world consumed as much paper today as the average American (341 kilograms a year), the world would be using nearly seven times as much paper. And by 2050 it would need more than 11 times as much. If, on the other hand, paper use stabilized at today's global average—50 kilograms a year per person—and if it were distributed more equitably, paper consumption in 2050 could be held to 1.7 times today's level.[125]

Lowering waste and consumption would benefit forests and economies without sacrificing quality of life.

Real options exist to reduce the amount of wood consumed for fuel. Shifting the industrial and urban sectors—who use most of the wood fuel—to methane, propane, and clean, renewable energy sources could greatly reduce the pressures on the forests and improve air quality as well.[126]

Incorporating the full costs of management and production into the price of forest and agricultural products would encourage more judicious use by producers and consumers. To do this, many costly incentives and subsidies need to be eliminated, such as below-cost timber sales, give-away forest concessions, and subsidized forest conversion. Other policies—such as dissecting forests with roads that encourage forest conversion and granting title to those who clear the land—also need serious reform in order to ensure that they do not promote forest degradation.[127]

Much forest mismanagement, waste, and overconsumption result from the fact that only a fraction of forest goods are counted when they enter the marketplace, and that forest services—the life-support systems—are not counted at all. As environmental consultant Norman Myers puts it, "our tools of economic analysis are far from able to apprehend, let alone comprehend, the entire range of values implicit in forests." The profit from deforesting land is counted as an addition to the national economy, but the depletion of timber, fisheries, or watersheds and the loss of climate services are not subtracted. These errors send misleading economic signals to decisionmakers at all levels.[128]

In the last few years, a new breed of economists—ecological economists—have been offering methods to correct faulty economic signals. Alternative methods for calculating the benefits from forests and nature are being developed. When the depletion of petroleum, timber, and soil were subtracted from Indonesia's GDP, for instance, the real economic growth rate from 1971–84 was 4 percent, not the rosy 7.1 percent that the GDP estimated. Capturing the value of a forest's ecological services to support sustainable rural development represents an important step forward in improving decisionmaking.[129]

As described earlier, the financial benefits from forest exploitation frequently go to private individuals or entities, while the economic, social, and environmental losses are distributed across society. When a small segment of society profits from unsustainable forest exploitation and the rest (and future generations) bear the costs, there is little economic incentive for those exploiting a resource to use it judiciously or in a manner that maximizes public good.[130]

A proven way to reconnect the costs and benefits of forest management is by returning—or devolving—control of forests to communities. Community participation in planning and management can improve the sustainability of the forests and the quality of life of people living in them and nearby. In India, for example, when the state assumed control over forests from local communities over a century ago, they removed the only successful safeguard against overexploitation, and the condition of forests declined. After the policy was modified in the late 1980s, thousands of communities regained control over state forestlands.

Communities now protect and control—and benefit from— the forests that they manage and rehabilitate. In Indonesia, likewise, reinstating customary rights could help reverse the degradation and poverty caused by the last few decades of state and industrial control over the forests. Community forest control can also improve the quality of forests and communities in industrialized nations in places like British Columbia, and many groups are trying to make it happen.[131]

There is much room for improvement of domestic laws and policies governing forests. Not only eliminating subsidies and reforming tenure policies, but also improving revenue collection from public lands are important elements. So, too, is better enforcement of existing laws, including preventing illegal logging and trade. Investments in strengthening forest law enforcement will provide short- and long-term benefits. These changes make good economic and ecological sense, but they can come about only if powerful interests are persuaded or compelled to accept them. Future progress will be difficult if the current breakdown in the rule of law governing forests and forest products is allowed to continue.

Although much of the action on forests needs to take place at the national and local level, there are also important roles for international agreements, institutions, and initiatives. Forests are a global issue. They cross political boundaries, as do their problems. And many of the services forests provide—such as storing carbon, regulating the climate, and sustaining biodiversity—are shared globally as well.

Governments need to renew the commitments made in Rio de Janeiro in 1992 and to accelerate action. In the years leading up to the 1992 Earth Summit, tropical forests were a major focus of international concern. When it came time to negotiate a binding forest convention, southern nations were concerned that northern governments would use a convention to impose controls on tropical forests that northerners were unwilling to accept at home—a tension that persists today. At the eleventh hour, a set of "Non-legally Binding...Forest Principles" that applies to all forests was adopted.[132]

Nations did agree to two legally binding instruments that provide significant opportunities for cooperation and meaningful action on forests—the Framework Convention on Climate Change and the Convention on Biological Diversity. The latter treaty, signed by 169 nations, has the conservation and sustainable and equitable use of biodiversity—including forests—as its mandate. Forests will be a major agenda item when the signatories meet in May 1998.[133]

Agenda 21—the plan of action that emerged from the Earth Summit—contains a chapter called "Combating Deforestation" that also provides guidance for action. Nations agreed to sustain the multiple roles of all types of forests, to enhance sustainable management and conservation, to rehabilitate degraded forests, to value and use forest goods and services more fully, and to improve the quality and availability of information about forests.[134]

Given the lack of progress on combating deforestation since Rio—indeed, the situation has grown worse—the United Nations set up an Intergovernmental Panel on Forests (IPF) in 1995. Its goal was to facilitate discussion by governments on a broad—some say too broad—range of issues and provide concrete recommendations for moving forward. A separate World Commission on Forests and Sustainable Development was also created, consisting of scientists, policymakers, and others.[135]

At the United Nations' five-year review of progress since the Earth Summit, a successor to the IPF was designated to implement its proposals for action and deal with issues left pending. After its first meeting in October 1997, the new Intergovernmental Forum on Forests urged nations to examine the underlying causes of deforestation and develop strategies to address them.[136]

One initiative still under consideration is a global forest convention. Ironically, a forest convention could delay action. Negotiating and ratifying an international treaty can take a decade, and more years can elapse before substantive action begins. With few exceptions, governments have been unwilling to accept international agreements that have "teeth," because they are usually most responsive to pressures from interests that benefit from the status quo and because of their concerns about national sovereignty. Thus, it is likely that a forest convention would formalize weak, non-binding standards. Not coincidentally, many of the nations that now support a forest convention have powerful timber industries. Given the political realities and the urgency of the forest problem, the most effective course of action is to use existing mechanisms and legal instruments, such as the biodiversity and climate change conventions.[137]

The Framework Convention on Climate Change also has the potential to help cultivate a new relationship with forests. Under the Kyoto Protocol negotiated in late 1997, nations committed to reducing their greenhouse gas emissions must include in their calculations the changes to their carbon stock resulting from "afforestation, deforestation, and reforestation." While it could offer an important incentive for countries to better recognize the value of forests and provide an avenue for poorer nations to receive some international assistance, the protocol has yet to determine how, or whether, tree harvesting and replanting are included within these three activities. This potential loophole—opened by large timber-producing nations—could allow a nation to exclude the carbon emitted from harvesting a forest, yet get credit for the carbon absorbed by the subsequent regrowth. If nations do not get credits for maintaining existing forest stocks or debits for harvesting, a potentially powerful disincentive for forest conservation may be created. Signatories to the protocol will have an opportunity to close this loophole when they meet again in late 1998.[138]

There are also opportunities for international cooperation on forests in regional trade agreements and forums. To date, many of these trade alliances have been driving forest destruction, by easing trade barriers and making environmental considerations secondary to, or illegal under, trade laws. In the future, they could be used instead to secure a better future for their economies and environments. Existing trade treaties such as the International Tropical

Timber Agreement (ITTO), for example, could be reformed to cover the entire timber trade, not just tropical timber—a step that the parties failed to take when it was renegotiated in 1994. Likewise, the laudable goals of the ITTO's Guidelines for the Sustainable Management of Natural Tropical Forests by the year 2000 could be expanded to apply to the temperate and boreal forest product trade, and be made binding. The Convention on International Trade in Endangered Species of Wild Flora and Fauna (CITES) has had some success in halting the decline of a few listed species (such as elephants) by restricting or prohibiting their trade, but the record for tree species has not been as good.[139]

Currently, there are no legal mechanisms that allow nations to block the import of timber that has been illegally felled and exported. In fact, changes to the World Trade Organization have made this even more difficult. Modifying CITES to cover more timber species or creating new laws to cover these concerns would help eliminate some of the illegal timber trade.[140]

There is also no mechanism that allows compensation to people or nations that have suffered damage from environmental mismanagement (such as landslides caused by deforestation or smoke from forest burning in neighboring countries). One suggested mechanism for arbitrating global environmental problems is to establish an International Court for the Environment along the lines of the International Court of Justice or the International Court of Human Rights.[141]

Nor is there a way to compensate or reward communities or nations for maintaining intact forests that provide a host of globally important ecosystem services. At the 1992 Earth Summit, the G-7 nations pledged $1.5 billion in financial and technical aid for environmental protection in Brazil. So far, only $20 million has been forthcoming. The Global Environmental Facility was set up in 1991 by the World Bank, the U.N. Development Programme, and the U.N. Environment Programme as a way to provide extra funding for projects that have global environmental benefits. So far, about $1.6 billion has been allocated to help developing nations and nations in transition undertake such efforts under global treaties on climate, ozone, and biodiversity. While potentially contentious, compensation mechanisms and financial transfers may be even more crucial in the future.[142]

International lending and donor agencies also have a role to play by ensuring that their loans and grants encourage positive reforms and sustainable practices rather than corruption and deforestation. So, too, do private investors who are now responsible for the majority of financial transfers. Loans for dams, road building, agriculture, and resettlement schemes are examples of projects that contribute to deforestation. On the positive side, the World Bank announced in 1997 that it will help client nations meet the goals of having 10 percent of each forest type in protected areas and expanding the area under certified sustainable forest management by 200 million hectares by 2005. It has also begun a review of its 1991 Forest Sector Policy. Recently, the United Nations, the International Monetary Fund, and the World Bank made their future aid to Cambodia conditional on reforming and adhering to national forest laws and not violating the laws of neighboring nations. At the same time, however, the World Bank is also considering reversing course on one part of its policy by allowing logging in primary tropical forests.[143]

The Convention on Climate Change has the potential to help forests.

The Asian currency crisis provides a window of opportunity for parties within nations to bring about some positive reforms, with the help of international donors. The IMF rescue package to Indonesia during the 1997–98 currency crisis, for example, prescribed a number of reforms to the nation's economy such as banking and subsidy reform, dismantling of monopolies (including some owned by the president's family and associates), and greater transparency in public sector activities and finances. Many observers are calling for such international financial packages to contain more explicit protection of the environment.[144]

We have the opportunity and the know-how to cultivate a new relationship with the world's forests, one that will reverse their decline, improve people's quality of life, and ensure that future generations inherit healthy forests. Whether this relationship develops fast enough will depend on who wins the fierce competition now under way—between the powerful supporters of the status quo racing to harvest the remaining forests before someone else does and the growing ranks of environmentalists, scientists, local people, and business and government leaders pressing for a viable alternative. Whether or not the bystanders to this competition recognize its urgency and throw their support to a new relationship with the forests in time will determine the outcome.

Notes

1. United Nations Food and Agriculture Organization (FAO), *State of the World's Forests 1997* (Oxford, U.K.: 1997).

2. Nigel Dudley, "The Year the World Caught Fire," (Gland, Switzerland: World Wide Fund For Nature, December 1997).

3. Robin Broad, "The Political Economy of Natural Resources: Case Studies of the Indonesian and Philippine Forest Sectors," *The Journal of Developing Areas*, April 1995.

4. Dudley, op. cit. note 2.

5. FAO, "Importance of NWFP," <http://www.fao.org/WAICENT/FAOINFO/FORESTRY/NWFP>, viewed 11 February 1998; Ganges from George Ledec and Robert Goodland, *Wildlands: Their Protection and Management in Economic Development* (Washington, DC: World Bank, 1988); Malaysia from David W. Pearce and Dominic Moran, *The Economic Value of Biological Diversity* (London: Earthscan Publications, 1994).

6. Sander Thoenes, "In Asia's Big Haze, Man Battles Man-Made Disaster," *Christian Science Monitor*, 28 October 1997.

7. Belize from "Mayan Homeland in Belize Rainforest Under Siege by Malaysian Loggers," *Fourth World Bulletin*, spring/summer 1996, reprinted from *Belize Times*, 5 November 1995; Indonesia from Charles Barber, Nels C. Johnson, and Emmy Hafild, *Breaking the Logjam: Obstacles to Forest Policy Reform in Indonesia and the United States* (Washington, DC: World Resources Institute (WRI), 1994); U.S. from Jim Jontz, "Forest Service Indictment: A Mountain of Evidence," in Sierra Club, *Stewardship or Stumps? National Forests at the Crossroads* (Washington, DC: June 1997); Randal O'Toole, "Reforming a Demoralized Agency: Saving National Forests," *Different Drummer*, vol. 3, no. 4 (1997); "National Forest Timber Sale Receipts and Costs in 1995," *Different Drummer*, vol. 3, no. 4 (1997); Paul Roberts, "The Federal Chain-saw Massacre," *Harper's Magazine*, June 1997.

8. Trade from FAO, op. cit. note 1; Brazil from Secretaria de Assuntos Estratégicos, *Grupo de Trabalho sobre Política Florestal: A Exploraçào Madeireira na Amazônia. Relatório*. Brasília, 8 April 1997, cited in Stephan Schwartzman, "Fires in the Amazon—An Analysis of NOAA-12 Satellite Data 1996–1997," factsheet (Washington, DC: Environmental Defense Fund (EDF), 23 September 1997).

9. Wood demand from FAO, FAOSTAT Statistics Database, <http://apps.fao.org/>, viewed January 1998; population from U.S. Bureau of the Census, *International Data Base*, <http://www.census.gov/cgi-bin/ipc>, viewed 18 December 1997; projected demand from FAO, op. cit. note 1.

10. Forty percent derived from estimated world industrial wood harvest of 1,530 million cubic meters in 1993, cited in Institute for Environment and Development (IIED), *Towards a Sustainable Paper Cycle* (London: 1996); reported 1993 wood pulp production of 618 million cubic meters from Wood Resources International Ltd., "Fiber Sourcing Analysis for the Global Pulp and Paper Industry" (London: IIED, September 1996); 130 million trees from Environmental Investigation Agency (EIA), *Corporate Power, Corruption and the Destruction of the World's Forests* (Washington, DC: September 1996).

11. Forty percent recovered from Pulp and Paper International (PPI), *International Fact and Price Book 1997* (San Francisco, CA: Miller Freeman, 1996); contribution of recovered paper to total fiber calculated from FAO, op. cit. note 9.

12. Philip M. Fearnside, "Environmental Services as a Strategy for Sustainable Development in Rural Amazonia," *Ecological Economics*, no. 20, 1997; Indonesia example from H.J. Ruitenbeek, Mangrove Management, *An Economic Analysis of Management Options with a Focus on Bintuni Bay, Irian Jaya*, Environmental Reports No. 8 (Gabriola Island, BC, Canada: Environmental Management Project, 1992).

13. Janet N. Abramovitz, "Valuing Nature's Services," in Lester R. Brown et al., *State of the World 1997* (New York: W.W. Norton & Company, 1997); Robert Costanza et al., "The Value of the World's Ecosystem Services and Natural Capital," *Nature*, 15 May 1997; Fearnside, op. cit note 12; Norman Myers, "The World's Forests and Their Ecosystem Services," in Gretchen C. Daily, ed., *Nature's Services* (Washington, DC: Island Press, 1997).

14. Value of global trade from FAO, op. cit. note 5, and FAO, op. cit. note 9; employment data from Ravinder Kaur, "Women in Forestry in India," in Background Paper for World Bank Women and Development in India Review (Washington, DC: World Bank, 1990); Indonesia from Theodore Panayotou and Peter S. Ashton, *Not by Timber Alone: Economics and Ecology for Sustaining Tropical Forests* (Washington, DC: Island Press, 1992); Ricardo A. Godoy et al., "A Method for the Economic Valuation of Non-Timber Forest Products," *Economic Botany*, vol. 47, no. 3 (1993); Julian A. Lampietti and John A. Dixon, *To See the Forest for the Trees: A Guide to Non-Timber Forest Benefits*, Environment Department Paper No. 013 (Washington, DC: World Bank, 1995).

15. Carol Ireson, "Women's Forest Work in Laos," *Society and Natural Resources*, vol. 4, 1991; Ghana example in Florence Addo et al., "The Economic Contribution of Women and Protected Areas: Ghana and the Bushmeat Trade," paper presented at the IV World Parks Congress on National Parks and Protected Areas, Caracas, Venezuela, 10–21 February 1992; Julia Falconer, *The Major Significance of 'Minor' Forest Products: The Local Use and Value of Forests in the West African Humid Forest Zone* (Rome: FAO, 1990).

16. Cork from FAO, *Non-Wood News*, no. 4, <http://www.fao.org/
WAICENT/FAOINFO/FORESTRY>, viewed 11 February 1998; British
Columbia Ministry of Forests (MOF), "Botanical Forest Products in British
Columbia: An Overview," April 1995, <http://www.for.gov.bc.ca>, viewed 5
February 1998; W.E. Schlosser et al., "Economic and Marketing Implications
of Special Forest Products Harvest in the Coastal Pacific Northwest,"
Western Journal of Applied Forestry, vol. 6, 1991.

17. Women in the NTFP economy from Kaur, op. cit. note 14, from D.D.
Tewari, "Developing and Sustaining Non-timber Forest Products: Policy
Issues and Concerns with Special Reference to India," *Journal of World Forest
Resource Management*, vol. 7, 1994, and from Jeffrey Campbell, *Case Studies
in Forest-Based Small Scale Enterprises* (Bangkok: FAO, 1991); rubber tappers
from Susanna Hecht, "Sustainable Extraction in Amazonia," in Lea M.
Borkenhagen and Janet N. Abramovitz, eds., *Proceedings of the International
Conference on Women and Biodiversity* (Washington, DC: Committee on
Women and Biodiversity and WRI, 1993), and Constance E. Campbell, "On
the Front Lines but Struggling for a Voice: Women in the Rubber Tappers'
Defense of the Amazon Forest," *The Ecologist*, March/April 1997; India
example from Kaur, op. cit. note 14; quote from Madhav Gadgil,
"Biodiversity and India's Degraded Lands," *Ambio*, May 1993.

18. Amazon fish species from Peter B. Moyle and Robert A. Leidy, "Loss of
Biodiversity in Aquatic Ecosystems: Evidence from Fish Faunas," in P.L.
Fiedler and S.K. Jain, eds., *Conservation Biology: The Theory and Practice of
Nature Conservation, Preservation, and Management* (New York: Chapman and
Hall, 1992); Paul Ehrlich and Anne Ehrlich, "The Value of Biodiversity,"
Ambio, May 1992; Myers, op. cit. note 13.

19. Kanta Kumari, "Sustainable Forest Management: Myth or Reality?
Exploring the Prospects for Malaysia," *Ambio*, November 1996.

20. Norman Myers, "The World's Forests: Problems and Potentials,"
Environmental Conservation, vol. 23, no. 2 (1996); David Pimentel et al.,
"Environmental and Economic Costs of Soil Erosion and Conservation
Benefits," *Science*, 24 February 1995; long-term impact of logging on
streams from Christopher A. Frissell, *A New Strategy for Watershed Restoration
and Recovery of Pacific Salmon in the Pacific Northwest* (Corvallis, OR: Pacific
Rivers Council, 1993).

21. Deforestation in Ganges river valley from Ledec and Goodland, op.
cit. note 5; British Columbia cited in Nigel Dudley, Jean-Paul Jeanrenaud,
and Francis Sullivan, *Bad Harvest? The Timber Trade and the Degradation of
the World's Forests* (London: Earthscan Publications, 1995); Pacific
Northwest from William Weaver and Danny K. Hagans, "Aerial
Reconnaissance Evaluation of 1996 Storm Effects on Upland Mountainous
Watersheds of Oregon and Southern Washington: Wildland Response to
the February 1996 Storm and Flood in the Oregon and Washington
Cascades and Oregon Coast Range Mountains," paper prepared for Pacific

Rivers Council, Eugene, OR (Arcata, CA: Pacific Watershed Associates, May 1996); "A Tale of Two Cities—and Their Drinking Water," in Sierra Club, op. cit. note 7; Romain Cooper, "Floods in the Forest," *Headwaters' Forest News*, spring 1997; David Bayles, "Logging and Landslides," *New York Times*, 19 February 1997; William Claiborne, "When a Verdant Forest Turns Ugly: 8 Oregon Deaths Blamed on Mud Sliding Down Clear-Cut Hillsides," *Washington Post*, 18 December 1996.

22. Ehrlich and Ehrlich, op. cit. note 18, Myers, op. cit. note 20; carbon from Asia fires from Thoenes, op. cit. note 6; Chris Bright, "Tracking the Ecology of Climate Change," in Brown, et al., op. cit. note 13.

23. Richard A. Houghton, "Converting Terrestrial Ecosystems from Sources to Sinks of Carbon," *Ambio*, 4 June 1996; Sandra Brown et al., "Management of Forests for Mitigation of Greenhouse Gas Emissions," in Robert T. Watson et al., eds., *Climate Change 1995: Impacts, Adaptations and Mitigation of Climate Change: Scientific-Technical Analyses: Contribution of Working Group II to the Second Assessment Report of the Intergovernmental Panel on Climate Change* (New York: Cambridge University Press, 1996).

24. Oak from Caspar Henderson "Right Climate for Change," *Financial Times*, 6 August 1997; carbon storage in forest types from Paige Brown, Bruce Cabarle, and Robert Livernash, *Carbon Counts: Estimating Climate Change Mitigation in Forestry Projects* (Washington, DC: WRI, September 1997); U.S. from Gerry Gray, "Carbon Debt: We All Have One," *American Forests*, summer 1996.

25. Brown et al., op. cit. note 23; New Zealand from Anne Simon Moffat, "Resurgent Forests Can Be Greenhouse Sponges," *Science*, 18 July 1997.

26. Houghton, op. cit. note 23; Brown et al., op. cit. note 23.

27. Brown, Cabarle, and Livernash, op. cit. note 24; Moffat, op. cit. note 25; reduced impact of logging from Michele A. Pinard and Francis E. Putz, "Retaining Forest Biomass by Reducing Logging Damage," *Biotropica*, vol. 28, no. 3 (1996); carbon valuation from Neil Adger et al., "Total Economic Value of Mexican Forests," *Ambio*, August 1995, from Jan Bojo, "Economic Valuation of Indigenous Woodlands in Zimbabwe," in P. Bradley and K. McNamara, eds., *Living with Trees: A Future for Social Forestry in Zimbabwe* (Washington, DC: World Bank, 1993), from Katrina Brown and David W. Pearce, "The Economic Value of Non-marketed Benefits of Tropical Forests: Carbon Storage," in J. Weiss, ed., *The Economics of Project Appraisal and the Environment* (London: Edward Elgar, 1994), from David Pearce, "Deforesting the Amazon: Toward an Economic Solution," *Ecodecision*, first quarter 1991, and from Pearce and Moran, op. cit. note 5.

28. Costanza et al., op. cit. note 13; gross world product from International Monetary Fund (IMF), *World Economic Outlook, October 1996* (Washington, DC: 1996).

29. FAO, op. cit. note 1.

30. Dirk Bryant, Daniel Nielsen, and Laura Tangley, *The Last Frontier Forests: Ecosystems and Economies on the Edge* (Washington, DC: WRI, 1997); 200 million from FAO, op. cit. note 1. (This loss was partially offset by an increase of 20 million hectares in developed countries for a net decrease in forest area of 180 million hectares over the 15-year period.) Table 1 from Bryant, Nielson, and Tangley, op. cit. this note; net annual change from FAO, op. cit. note 1.

31. Ibid. Figure 2 from ibid.

32. Ibid.

33. Historical losses from Dudley, Jeanrenaud, and Sullivan, op. cit. note 21; losses since 1960 from WRI, *World Resources 1996–97* (New York: Oxford University Press, 1996).

34. Deforestation 1980–95 from FAO, op. cit. note 1 and WRI, *World Resources, 1994–95* (New York: Oxford University Press, 1994); tropical dry forests from Anil Agarwal, "Dark Truths and Lost Woods," *Down to Earth*, 15 June 1997; mangrove forests from Solon Barraclough and Andrea Finger-Stich, *Some Ecological and Social Implications of Commercial Shrimp Farming in Asia*, Discussion Paper 74 (Geneva: U.N. Research Institute for Social Development, March 1996); temperate rainforests in North America from Dominick DellaSala et al., "Protection and Independent Certification: A Shared Vision for North America's Diverse Forests," mapping analysis (Washington, DC: World Wildlife Fund–US and World Wildlife Fund–Canada, 1997), and from Conservation International, Ecotrust, and Pacific GIS, "Coastal Temperate Rain Forests of North America," map (Washington, DC, and Portland, OR: 1995).

35. Bryant, Nielsen, and Tangley, op. cit. note 30; Reed Noss, E.T. LaRoe III, and J.M. Scott, *Endangered Systems of the United States: A Preliminary Assessment of Loss and Degradation* (Washington, DC: U.S. Department of the Interior, National Biological Service, 1995); DellaSala et al., op. cit. note 34.

36. Dudley, Jeanrenaud, and Sullivan, op. cit. note 21.

37. FAO, op. cit. note 1. FAO records plantation area for developing countries only. However, they report "rough" estimates of plantation cover in industrial countries as 45–60 million hectares in 1980 and 80–100 million hectares in 1995. Recorded figures for developing countries were 40.2 million hectares in 1980 and 81.2 million in 1995—thus, a rough approximation of global cover would be 85–100 million hectares in 1980 and 161–181 million hectares in 1995. FAO forest plantation cover estimates do not include cover of rubber, coconut, and palm oil plantations; however, they estimate that in 1990 there were over 14 million hectares of these types of plantations, likely a very conservative estimate. Portion for industrial use

from Michael D. Bazett, *Industrial Wood, Study No. 3, Shell/Worldwide Fund for Nature Tree Plantation Review* (London: Shell Petroleum Company and World Wildlife Fund (WWF) 1993), cited in Ricardo Carerre and Larry Lohmann, *Pulping the South, Industrial Tree Plantations and the World Paper Economy* (London: Zed Books, 1996).

38. Dudley, Jeanrenaud, and Sullivan, op. cit. note 21; New Zealand from Rowan Taylor and Ian Smith, *The State of New Zealand's Environment 1997* (Wellington: Ministry for the Environment and GP Publications, 1997).

39. U.N. Economic Commission for Europe, International Co-operative Programme on Assessment and Monitoring of Air Pollution Effects on Forests, "Forest Condition Report 1996 (Summary)," <http://www. dainet.de/bfh/icpfor/icpfor.htm>, viewed 23 October 1997.

40. Plans for FRA 2000 from Susan Braatz, Policy Analyst/Coordinator of *State of the World's Forests 1997*, FAO, e-mail to author, 9 February 1998.

41. FAO, op. cit. note 1; Dudley, Jeanrenaud, and Sullivan, op. cit. note 21; slash-and-burn farming from Consultative Group on International Agricultural Research (CGIAR), "Poor Farmers Could Destroy Half of Remaining Tropical Forest," press release, <http://www.worldbank.org/ html/cgiar/ press.forest.html>, viewed 23 October 1997.

42. Indonesia from A. Ruwindrijarto, Programme Coordinator, Telapak, presentation at Global Forest Watch Workshop, 15 January 1998, Washington, DC; Brazil from Diana Jean Schemo, "Data Show Recent Burning of Amazon Is Worst Ever," *New York Times*, 27 January 1998; Stephan Schwartzman "Fires in the Amazon—An Analysis of NOAA-12 Satellite Data 1996–1997," factsheet (Washington, DC: EDF, 1 December 1997).

43. FAO, op. cit. note 9. Table 2 from ibid.

44. Ibid. Table 3 from FAO, op. cit. note 1; population from U. S. Bureau of the Census, op. cit. note 9, using FAO lists of developed and developing countries; consumption percentages calculated from FAO, op. cit. note 9.

45. Emmanuel N. Chidumayo, "Woodfuel and Deforestation in Southern Africa—A Misconceived Association," *Renewable Energy for Development*, July 1997; Brazil tobacco example from Beauty Lupiya, "All for Smoke," *Down to Earth*, 15 November 1997.

46. Dudley, Jeanrenaud, and Sullivan, op. cit. note 21; FAO, op. cit. note 1.

47. Ibid. Figure 2 from FAO, op. cit. note 1.

48. FAO, op. cit. note 9; FAO, op. cit. note 1; population from U.S. Bureau of the Census, op. cit. note 9.

49. FAO, op. cit. note 9; Philippines from Broad, op. cit. note 3; Japan imports from FAO, *FAO Forest Products Yearbook 1983–1994* (Rome: 1996) and FAO, op. cit. note 9; Russia's drop in production from FAO, op. cit. note 9.

50. International forest products trade volume and value from FAO, op. cit. note 9 (value reported is for 1994 trade). The value of trade for all raw materials and manufactured goods in 1994 was $4,107 billion, according to the *World Trade Organization Annual Report 1996* (Geneva: 1996); importance of temperate and boreal forests also from James McIntire, ed., *The New Eco-Order: Economic and Ecological Linkages of the World's Temperate and Boreal Forest Resources* (Seattle: Northwest Policy Center, University of Washington, 1995). Table 4 from FAO, op. cit. note 9; inflation conversion factors from Robert Sahr, Oregon State University, Department of Political Science <http://www.osu.orst.edu/Dept/pol_sci>, viewed 6 March 1998. Table 5 from FAO, op. cit. note 1.

51. Ghana from Friends of the Earth (FOE) International, *Cut and Run: Illegal Logging and Timber Trade in Four Tropical Countries* (Amsterdam: March 1997); Brazil from Secretaria de Assuntos Estratégicos, op. cit. note 8.

52. Production expansion of industrial roundwood and plywood from FAO, op. cit. note 9; FAO, op. cit. note 49. Forest loss in Indonesia, Brazil, and Malaysia from WRI, op. cit. note 34.

53. Asian growth from FAO, op. cit. note 1.

54. 3.3 percent from FAO, op. cit. note 1; increase since 1961 from FAO, op. cit. note 9; 40 percent calculated from IIED, op. cit. note 10, and Wood Resources International, op. cit. note 10; projection to 50 percent from Nigel Dudley, Sue Stolton, and Jean-Paul Jeanrenaud, *Pulp Fact, The Environmental and Social Impacts of the Pulp and Paper Industry* (Gland, Switzerland: WWF, 1995); paper products from PPI, cited in IIED, op. cit. note 10.

55. Figure 3 from FAO, op. cit. note 9, and U.S. Bureau of the Census, op. cit. note 9. Figure 4 from FAO, op. cit. note 9, and U.S. Bureau of the Census, op. cit. note 9, projections to 2010 from IIED, op. cit. note 10.

56. Fiber sources from IIED, op. cit. note 10; share of recovered paper in fiber furnish from FAO, op. cit. note 9.

57. Recovery rates calculated with data provided in PPI, op. cit. note 11; U.S. industry goal from Barry Polsky, Director of Media Relations, American Forest and Paper Association, discussion with author, 30 September 1997; municipal solid waste from Franklin Associates, Ltd., *Characterization of Municipal Solid Waste in the United States: 1996 Update*, report prepared for the U.S. Environmental Protection Agency (EPA), Municipal and Industrial Solid Waste Division, Office of Solid Waste, June 1997.

58. U.K. paper discarded, U.S. use, German survey, and lumber for crates and pallets from EIA, op. cit. note 10.

59. Renewable energy from Christopher Flavin and Seth Dunn "Responding to the Threat of Climate Change," in Lester R. Brown et al., *State of the World 1998* (New York: W.W. Norton & Company, 1997).

60. FAO, op. cit. note 1; collateral damage from Government of Indonesia and FAO in Barber, Johnson, and Hafild, op. cit. note 7, and Pinard and Putz, op. cit. note 27.

61. Sorting and grading from the Sustainable Forestry Working Group (SFWG), *Sustaining Profits and Forests: The Business of Sustainable Forestry* (Chicago: John D. and Catherine T. MacArthur Foundation, 1997), and Jim Smith, "Bringing Ecoforestry to the BC Forest Service," *Global Biodiversity*, vol. 7, no. 2, fall 1997; wood used in making shipping crates from Catherine Mater, Mater Engineering, presentation at WWF Forests for Life Conference, San Francisco, CA, 8–10 May 1997.

62. U.S. sawmill wastes used for pulp and fuel from Maureen Smith, *The U.S. Paper Industry and Sustainable Production* (Cambridge, MA: MIT Press, 1997); Canada from British Columbia Ministry of Forests (MOF), *1994 Forest, Range and Recreation Resource Analysis*, <http://www.for.gov.bc.ca/pab/publctns/frrra>, viewed 5 February 1998; Brazil from Christopher Uhl et al., "Natural Resource Management in the Brazilian Amazon: An Integrated Research Approach," *Bioscience*, March 1997; FAO, op. cit. note 1.

63. Abramovitz, op. cit. note 13; Reed Noss and Allen Cooperrider, *Saving Nature's Legacy* (Washington, DC: Island Press, 1994); Myers, op. cit. notes 13 and 20.

64. Hunters from Michael McRae, "Road Kill in Cameroon," *Natural History*, February 1997; Leandro V. Ferreira and William F. Laurance, "Effects of Forest Fragmentation on Mortality and Damage of Selected Trees in Central Amazonia," *Conservation Biology*, June 1997; William F. Laurance and R.O. Bierregaard, Jr., eds., *Tropical Forest Remnants: Ecology, Management and Conservation of Fragmented Communities* (Chicago: University of Chicago Press, 1997).

65. Road density and length (reported as 380,000 miles) in U.S. national forests from Carey Goldberg, "Quiet Roads Bringing Thundering Protests; Congress to Battle Over Who Pays to Get to National Forest Trees," *New York Times*, 23 May 1997; L. Potter, "Forest Degradation, Deforestation and Reforestation in Kalimantan: Towards a Sustainable Land Use?" paper presented at the conference on "Interactions of People and Forests in Kalimantan," New York Botanical Garden, 21–23 June 1991, cited in Barber, Johnson, and Hafild, op. cit. note 7; Roads in British Columbia MOF, op. cit. note 62; percent of roads allowed from Scientific Panels for Sustainable Forest Practice in Clayoquot Sound, "Executive Summaries of Scientific

Panel Reports" (Victoria, BC: 1995), <http://www.interchg.ubc.ca/cacb/panel>, viewed 4 February 1998.

66. Highway to Brasília from Emilio F. Moran, "Deforestation in the Brazilian Amazon," in Leslie E. Sponsel, Thomas N. Headland, and Robert C. Bailey, eds., *Tropical Deforestation: The Human Dimension* (New York: Columbia University Press, 1996); amount of forest cleared for pasture from Dennis Mahar, cited in ibid.; increase in timber production from FOE, op. cit. note 51; deforested area in the Amazon in 1975–88 from David Skole and Compton Tucker, cited in Moran, op. cit. this note; current deforestation from Secretaria de Assuntos Estratégicos, op. cit. note 8; agricultural expansion from Atossa Soltani and Tracey Osborne, *Arteries of Global Trade, Consequences for Amazonia* (Malibu, CA: Amazon Watch, April 1997); area of Amazon basin (5,870,000 km^2) and Brazilian Amazon (3,715,000 km^2) from Peter Gleick, ed., *Water in Crisis: A Guide to the World's Freshwater Resources* (New York: Oxford University Press, 1993).

67. Soltani and Osborne, op. cit. note 66; Fundaçao Centro Brasileiro de Referencia e Apoio Cultural (CEBRAC) and WWF, "Parana-Paraguay Waterway: Who Pays the Bill?" Executive Summary, CEBRAC, Brasília, September 1994.

68. Michael Christie, "The Amazon Is Burning Again, Officials Say," Reuters, 3 October 1997; Schwartzman, op. cit. note 42; Dudley, op. cit. note 2; Dominick A. DellaSala et al., "Forest Health: Moving Beyond Rhetoric to Restore Healthy Landscapes in the Inland Northwest," *Wildlife Society Bulletin,* vol. 23, no. 3, 1995.

69. Robert G. Kaiser, "Forests of Borneo Going Up in Smoke," *Washington Post,* 7 September 1997; "Rain Forests on Fire: Conservation Consequences," WWF, <http://www.worldwildlife.org>, viewed 26 September 1997; Lewa Pardomuan, "Officials: Indonesian Forest Fires Spreading," Reuters, 13 October 1997; "Forest Fires Multiply," *Indonesia Times,* 15 October 1997; carbon dioxide estimate from Thoenes, op. cit. note 6; 1997 financial loss estimates from Dudley, op. cit. note 2.

70. Dudley, Jeanrenaud, and Sullivan, op. cit. note 21; Ana Toni, *Logging the Planet, Asian Companies Report* (Amsterdam: Greenpeace International, May 1997); FAO, op. cit. note 1.

71. Jonathon Friedland and Raphael Pura, "Log Heaven: Troubled at Home, Asian Timber Firms Set Sights on the Amazon," *Wall Street Journal,* 11 November 1996; Soltani and Osborne, op. cit. note 66; Carlos Sergio Figueiredo Tautz, "The Asian Invasion: Asian Multinationals Come to the Amazon," *Multinational Monitor,* September 1997; PNG from Timothy M. Ito and Margaret Loftus, "Cutting and Dealing: Asian Loggers Target the World's Remaining Rain Forests," *U.S. News and World Report,* 10 March 1997, and Dudley, Jeanrenaud, and Sullivan, op. cit. note 21.

72. These trade agreements include the General Agreement on Tariffs and Trade (GATT), the North American Free Trade Agreement (NAFTA), the European Union (EU), the Southern Cone Common Market (MERCOSUR), and the Association of South East Asian Nations (ASEAN); Nigel Sizer and Richard Rice, *Backs to the Wall in Suriname: Forest Policy in a Country in Crisis* (Washington, DC: WRI, April 1995); Nigel Sizer, *Profit Without Plunder: Reaping Revenue from Guyana's Tropical Forests Without Destroying Them* (Washington, DC: WRI, September 1996); Soltani and Osborne, op. cit. note 67.

73. Solomon Islands from EIA, op. cit. note 10; Suriname from Sizer and Rice, op. cit. note 72, and Nigel Sizer, WRI, telephone conversation with author, 8 February 1998; Belize from *Fourth World Bulletin*, op. cit. note 7; NGO Network for Forest Conservation in Indonesia in association with International Fund for Animal Welfare (SKEPHI), *Asian Forestry Incursions, Indonesian Logging in Surinam: Report on N.V. MUSA Indo-Surinam* (Jakarta: SKEPHI, 1996).

74. Natural Resources Canada, Canada Forest Service, *State of Canada's Forests 1995–1996* (Ottawa, ON: Natural Resources Canada, 1996); Barber, Johnson, and Hafild, op. cit. note 7.

75. Robert Repetto, *The Forest for the Trees? Government Policies and the Misuse of Forest Resources* (Washington, DC: WRI, 1988).

76. British Columbia MOF, "Stumpage: an information paper on timber pricing in British Columbia" (Victoria, BC: August 1996), <http://www.for.gov.bc.ca/revenue/timberp>, viewed 5 February 1998; Greenpeace Canada, *Broken Promises*, report produced in consultation with the Sierra Legal Defense Fund (SLDF) (Vancouver, BC: 1997); Indonesia from Barber, Johnson, and Hafild, op. cit. note 7.

77. Josh Newell and Emma Wilson, *The Russian Far East, Forests, Biodiversity Hotspots, and Industrial Developments* (Tokyo: FOE–Japan, 1996); World Bank, Agriculture, Industry and Finance Division, *Russian Federation Forest Policy Review, Promoting Sustainable Sector Development During Transition* (Washington, DC: 10 December 1996); Suriname from Sizer and Rice, op. cit. note 72; Guyana from Sizer, op. cit. note 72; Indonesia from Barber, Johnson, and Hafild, op. cit. note 7.

78. Jontz, op. cit. note 7; O'Toole, op. cit. note 7; Roberts, op. cit. note 7.

79. U. S. Forest Service, *The Forest Service Program for Forest and Rangeland Resources: A Long-Term Strategic Plan, Draft 1995 RPA Program*, October 1995, cited in Sierra Club, "Ending Timber Sales on National Forests: The Facts," report summary, undated.

80. Vernon forest from Greenpeace Canada, "Clearcut-free? Just did it," (Vancouver, BC: n.d.); SFWG, op. cit. note 61; Smith, op. cit. note 61; Silva

Forest Foundation, <http://www.silvafor.org/sertif/vernon.htm>, viewed 24 February 1998; U.S. example from Ken Muhlenfeld, Alabama Forestry Commission economist, and Cynthia West, U.S. Forest Service Hardwood Research Lab, Princeton, WV, cited in *Mobile Register,* 27 October 1996.

81. Tax concessions from Repetto, op. cit. note 75; trade from FAO, op. cit. notes 9 and 49; government plans to increase harvest levels from Barber, Johnson, and Hafild, op. cit. note 7.

82. Moran, op. cit. note 66; Repetto, op. cit. note 75; Leslie E. Sponsel, Robert C. Bailey, and Thomas N. Headland, "Anthropological Perspectives on the Causes, Consequences, and Solutions of Deforestation," in Sponsel, Headland, and Bailey, op. cit. note 66; taxes and credits for land in spite of reforms from Congressman Gilney Viana, *Initiatives in the Defense of the Amazon Rainforest* (Brasília: September 1996); infrastructure from Soltani and Osborne, op. cit. note 66; Angus Foster, "Brazil Seeks a 'Sustainable' Amazon," *Financial Times,* 19 April 1995.

83. Costs to government from Repetto, op. cit. note 75; failed resettlements from Nigel Dudley, forest researcher, Equilibrium, letter to author, 18 September 1997.

84. Thoenes, op. cit. note 6.

85. Myers, op. cit. note 20; Owen J. Lynch and Kirk Talbott, *Balancing Acts: Community-Based Forest Management and National Law in Asia and the Pacific* (Washington, DC: WRI, September 1995); Nancy L. Peluso and Christine Padoch, "Changing Resource Rights in Managed Forests of West Kalimantan," in Christine Padoch and Nancy L. Peluso, eds., *Borneo in Transition: People, Forests, Conservation, and Development* (New York: Oxford University Press, 1996); John W. Bruce and Louise Fortmann, "Why Land Tenure and Tree Tenure Matter: Some Fuel for Thought," in Louise Fortmann and John W. Bruce, eds., *Whose Trees? Proprietary Dimensions of Forestry* (Boulder, CO: Westview Press, 1988).

86. "Brazil Cuts Funds Used to Demarcate Indian Lands," *Baltimore Sun,* 11 October 1997; Guyana from Sizer, op. cit. note 72, and World Rainforest Movement (WRM), "Indigenous Peoples in Guyana Complain to Government about Multinational Miners and Loggers," WRM Forest Peoples' Program update (Moreton-in-Marsh, U.K.: 2 July 1997); Canada from "First Nations Perspectives Relating to Forest Practices Standards in Clayoquot Sound," Report 3 of the Scientific Panel for Sustainable Forest Practices in Clayoquot Sound (Victoria, BC: 1995), <http://www.interchg.ubc.ca/cacb/panel>, viewed 4 February 1998; FOE, op. cit. note 51; Mark Poffenberger and Betsy McGean, eds., *Village Voices, Forest Choices, Joint Forest Management in India* (Delhi: Oxford University Press, 1996); Madhav Gadgil, "India's Deforestation: Patterns and Processes," *Society and Natural Resources,* vol. 3, 1990; Shelton H. Davis and Alaka Wali, "Indigenous Land Tenure and Tropical Forest Management in Latin

America," *Ambio*, December 1994; Daniel Bromley, "Property Relations and Economic Development: The Other Land Reform," *World Development*, vol. 17, 1989; "Whose Common Future?" *The Ecologist* (entire issue), July/August 1992; Sponsel, Headland, and Bailey, op. cit. note 66; Marcus Colchester and Larry Lohmann, eds., *The Struggle for Land and the Fate of the Forests* (Penang, Malaysia: WRM, 1995); "Forest Gold," *The Economist*, 12 July 1997.

87.   Barber, Johnson, and Hafild, op. cit. note 7; Lynch and Talbott, op. cit. note 85.

88.   Broad, op. cit. note 3.

89.   Global Witness, "Just Deserts for Cambodia? Deforestation & the Co-Prime Ministers' Legacy to the Country," June 1997, <http://www.oneworld.org/globalwitness>, viewed 23 September 1997; Broad, op. cit. note 3.

90.   Broad, op. cit. note 3; Secretaria de Assuntos Estratégicos, op. cit. note 8; Barber, Johnson, and Hafild, op. cit. note 7; FOE, op. cit. note 51; Ito and Loftus, op. cit. note 71.

91.   Brazil from Secretaria de Assuntos Estratégicos, op. cit. note 8; IUCN and WWF, "Illegal Logging in Russian Forests," *Arborvitae*, August 1997.

92.   Papua New Guinea from Toni, op. cit. note 70; Ghana and Cameroon from FOE, op. cit. note 51.

93.   Global Witness, op. cit. note 89; Global Witness, "Illegal logging poses real threat to democratic process in Cambodia," press statement, 6 February 1998; "Cambodia: King Gives Backing to Report Calling for Overhaul of Government's Logging Policy," *International Environment Reporter*, 26 June 1996; value of exports, money to treasury from Daniel Pruzin, "Loggers Use Loophole to Decimate Cambodia's Disappearing Forests," *Christian Science Monitor*, 2 May 1997; Ted Bardacke, "Cambodia Failing to Curb Illegal Logging," *Financial Times*, 16 September 1997; importance of Tonle Sap from Janet N. Abramovitz, *Imperiled Waters, Impoverished Future: The Decline of Freshwater Ecosystems*, Worldwatch Paper 128 (Washington, DC: Worldwatch Institute, March 1996).

94.   IBAMA from Carlos Sergio Figueiredo Tautz, "The Asian Invasion: Asian Multinationals Come to the Amazon," *Multinational Monitor*, September 1997; Brazil laws from Diana Jean Schemo, "To Fight Outlaws, Brazil Opens Rain Forest to Loggers," *New York Times*, 21 July 1997; Diana Jean Schemo, "Brazil, Its Forests Besieged, Adds Teeth to Environmental Laws," *New York Times*, 29 January 1998; "Brazil Strengthens Protection of Natural Resources," *Baltimore Sun*, 13 February 1998; Suriname from Sizer and Rice, op. cit. note 72.

95. U.S. forests from Roberts, op. cit. note 7.

96. Indonesia reforestation fund from Broad, op. cit. note 3; "Environmental Group Again Loses Court Case on Alleged Diversion of Forest Funds," *International Environment Reporter*, 23 July 1997; fire-fighting from Australian Broadcasting Company, "IMF Says Indonesia was Unable to Use Special Fund Against Fires," 22 January 1998, <http://www.abc.net.au/ra/ newsrael>, viewed 23 January 1998.

97. Forest export values and distribution of crown land revenues by province from "National Forestry Database—Summary 1996," <http://www.nrcan.gc.ca/cfs/proj/iepb/nfdp/summary>, viewed 19 August 1997; rank of Canadian exports from FAO, op. cit. note 9; Cheri Burda et al., *Forests in Trust: Reforming British Columbia's Forest Tenure System for Ecosystem and Community Health* (Victoria, BC: University of Victoria, Eco-Research Chair of Environmental Law and Policy, July 1997); distribution of leases from Greenpeace Canada, op. cit. note 80.

98. BC Wild, "Overcut: British Columbia Forest Policy and the Liquidation of Old-Growth Forests," draft (Vancouver, BC: 1998, forthcoming); amount over sustainable yields from Tim Wilson, Mapping Director, BC Wild; analysis of BC Ministry of Forests Timber Supply Branch statistics, e-mail to author, 28 January 1998; R. Michael M'Gonigle, "Behind the Green Curtain," *Alternatives Journal*, fall 1997.

99. Greenpeace Canada, *Rainforest Ravagers* (Vancouver, BC: Greenpeace Canada, undated); two thirds and salmon stocks from Greenpeace Canada, op. cit. note 76; BC Wild, op. cit. note 98.

100. BC Forest Service, "Clayoquot Scientific Panel: Implications of Recommendations," <http://www.for.gov.bc.ca/het/clayoquot>, viewed 4 February 1998.

101. Scientific Panel for Sustainable Forest Practices in Clayoquot Sound, "Executive Summaries of Scientific Panel Reports," <http://www.interchg.ubc.ca/cacb/panel>, viewed 4 February 1997.

102. Province of British Columbia, "Government Adopts Clayoquot Scientific Report Moves to Implementation," press release, <http://www.for.gov.bc.ca/het/clayoquot>, viewed 4 February 1998.

103. SLDF, *Stream Protection Under the Code: The Destruction Continues*, report written on behalf of the Forest Caucus of the British Columbia Environmental Network (Vancouver, BC: February 1997); SLDF, *Wildlife at Risk*, report written on behalf of the Forest Caucus of the British Columbia Environmental Network (Vancouver, BC: April 1997); fines from Province of British Columbia, Ministry of Forests, "Annual Report of Compliance and Enforcement Statistics for the Forest Practices Code:

June 15, 1996–June 16, 1997," <http://www.for.gov.bc.ca.>, viewed 5 November 1997.

104. SLDF, "British Columbia's Clear Cut Code," factsheet (Vancouver, BC: November 1996); industry complaints and government response from Bernard Simon, "British Columbia Eases Logging Rules," *Financial Times,* 11 June 1997; "B.C. Environmental Regulations Said to Cost Forest Industry Share of U.S. Market," *International Environment Reporter,* 30 April 1997; B.G. Dunsworth and S.M. Northway, "Spatial Assessment of Habitat Supply and Harvest Values as a Means of Evaluating Conservation Strategies: A Case Study," in EFI Proceedings, *Assessment of Biodiversity for Improved Forest Planning* (Cambridge, MA: Kluwer Academic Publishers, 1997); Ministry of Forests, Province of British Columbia, "Important Changes to the Forest Practices Code," <http://www.for.gov.bc.ca/hfp/issues/amend/june09. htm>, viewed 8 August 1997; ecoforestry from M'Gonigle, op. cit. note 98, and Silva Forest Foundation, <http://www.silvafor.org>.

105. Timber production from FAO, op. cit. note 9; Canada cutting levels from BC Wild, op. cit. note 98; Share of tropics managed for sustained yield from Duncan Poore, "Conclusions," in Duncan Poore et al., *No Timber Without Trees* (London: Earthscan Publications, 1989); for initiatives in industry, see, for example, *Sustainable Forestry Initiative,* 2nd Annual Progress Report (Washington, DC: American Forest and Paper Association, 1997), and British Columbia Ministry of Forests, *Providing for the Future, Sustainable Forest Management in British Columbia* (Victoria, BC: March 1996).

106. Kathryn A. Kohm and Jerry F. Franklin, eds., *Creating a Forestry for the 21st Century: The Science of Ecosystem Management* (Washington, DC: Island Press, 1997); E. Thomas Tuchmann et al., *The Northwest Forest Plan: A Report to the President and Congress* (Portland, OR: U.S. Department of Agriculture, December 1996); Gregory H. Aplet et al., eds., *Defining Sustainable Forestry* (Washington, DC: Island Press, 1993); Narendra P. Sharma, ed., *Managing the World's Forests* (Dubuque, IA: Kendall/Hunt Publishing Company, 1992); international criteria and indicators from Richard G. Tarasofsky, *The International Forests Regime: Legal and Policy Issues* (Gland, Switzerland: World Conservation Union (IUCN) and WWF, December 1995).

107. Kohm and Franklin, op. cit. note 106; Herb Hammond, *Seeing the Forest Among the Trees* (Vancouver, BC: Polestar Press Ltd., 1991); Noss and Cooperrider, op. cit. note 63; Panayotou and Ashton, op. cit. note 14; Alan Drengson and Duncan Taylor, eds., *Ecoforestry* (Gabriola Island, BC: New Society Publishers, 1997); Duncan Poore, "The Sustainable Management of Tropical Forest: The Issues," in Simon Rietbergen, ed., *The Earthscan Reader in Tropical Forestry* (London: Earthscan Publications, 1993); WWF and IUCN, *Forests For Life* (Godalming, Surrey, U.K.: 1996).

108. Kohm and Franklin, op. cit. note 106; Noss and Cooperrider, op. cit. note 63; WWF and IUCN, op. cit. note 107; Panayotou and Ashton, op. cit. note 14; Hammond, op. cit. note 107; Ecotrust Canada, *Seeing the Ocean*

Through the Trees (Vancouver, BC: 1997).

109. Panayotou and Ashton, op. cit. note 14.

110. SFWG, op. cit. note 61; and idem, "Executive Summaries: The Cases" <http://www.sustainforests.org/execsum.htm>, viewed 6 February 1998; Vernon forest from Greenpeace Canada, op. cit. note 80; Silva Forest Foundation, op. cit. note 80.

111. "AssiDomän delivers Sweden's first FSC-approved wood," press release (Stockholm: AssiDomän, 12 November 1997), <http://www.asdo.se>, viewed 9 February 1998.

112. Kumari, op. cit. note 19; Ruitenbeek, op. cit. note 12; SFWG op. cit. note 61; Sizer, op. cit. note 72.

113. Catherine Mater, "Emerging Technologies for Sustainable Forestry," in SFWG, op. cit. 110.

114. Canadian Standards Association, "Standards for Canada's Forests," <http://www.sfms.com>, viewed 27 October 1997; Indonesian Ecolabelling Foundation in Jim Della-Giacoma, "Indonesian Forest Pact to Head Off Green Backlash," Reuters, 4 June 1996.

115. Forest Stewardship Council in WWF-UK, *World Wildlife Fund Guide to Forest Certification 1997*, Forests for Life Campaign (Godalming, Surrey, U.K.: 1997). Table 6 from WWF-UK, op. cit. this note; companies from Forest Stewardship Council (FSC), "Forests Certified by FSC-Accredited Certification Bodies," Document No. 5.3.3 (Oaxaca, Mexico: FSC, 28 February 1998).

116. Amount certified from WWF-UK, op. cit. note 115; current area certified from FSC, op. cit. note 115.

117. WWF-UK, op. cit. note 115, reports 2.4 billion pound sterling turnover for UK-1995 Plus groups; statements by Lennart Ahlgren, CEO AssiDomän, by Alan Knight, Environmental Policy Controller, B&Q, and by Nicholas Brett, Publishing Director of Radio Times, BBC Magazine, at WWF Forests for Life Conference, San Francisco, CA, 8–10 May 1997.

118. As of September 1997, there were buyers' groups in Australia, Brazil, Denmark, France, Germany, Ireland, Japan, Spain, Switzerland, the United Kingdom, and the United States; import and consumption data from FAO, op. cit. note 49; Japan certification from Dudley, op. cit. note 83.

119. WWF-UK, op. cit. note 115; Intergovernmental Seminar on Criteria and Indicators for Sustainable Forest Management, Background Document (Helsinki: Ministry of Agriculture and Forestry, June 1996).

120. Area certified as of February 1998 from FSC, op. cit. note 115; hectares to be certified by 2005 from Francis Sullivan, Director, WWF Forests for Life Campaign, briefing, Washington, DC, 11 November 1997; WWF, "World Bank and WWF Join Forces to Conserve Earth's Forests," press release (Washington, DC: 26 July 1997).

121. Repetto, op. cit. note 75; Myers, op. cit. note 20.

122. See, for example, WRI, "Blueprint for a Global Forest Watch," draft, (Washington, DC: 7 January 1998).

123. WWF-Netherlands, WWF, and World Conservation Monitoring Centre, "World Forest Map 1996" (Gland, Switzerland: WWF, 1997). Only 8 percent of tropical moist forests, 5 percent each of tropical dry and temperate needleleaf forests, 6 percent of temperate broadleaf forests, and 9 percent of mangroves have some protected status.

124. FAO, op. cit. note 1; Pinard and Putz, op. cit. note 27.

125. Paper consumption based on data from FAO, op. cit. note 9; population figures from U.S. Bureau of the Census, op. cit. note 9; population projections from U.N., *World Population Prospects: The 1996 Revision* (New York: forthcoming).

126. Fuelwood use from Chidumayo, op. cit. note 45; other energy sources from Flavin and Dunn, op. cit. note 59.

127. David Malin Roodman, "Reforming Subsidies," in Brown et al., op. cit. note 13.

128. Herman E. Daly and John B. Cobb, Jr., *For the Common Good* (Boston: Beacon Press, 1989); Clifford Cobb, Ted Halstead, and Jonathan Rowe, *Redefining Progress: The Genuine Progress Indicator, Summary of Data and Methodology* (San Francisco, CA: Redefining Progress, 1995); Repetto, op. cit. note 75; Robert Repetto et al., *Wasting Assets: Natural Resources in the National Income Accounts* (Washington, DC: WRI, 1989); Myers, op. cit. note 20.

129. Alternative measures of GDP and other methods of calculating benefits from Herman E. Daly, *Beyond Growth: The Economics of Sustainable Development* (Boston: Beacon Press, 1996); Cobb, Halstead, and Rowe, op. cit. 128; Repetto et al., op. cit. note 128; Panayotou and Ashton, op. cit. note 14; Wilfredo Cruz and Robert Repetto, *The Environmental Effects of Stabilization and Structural Adjustment Programs: The Philippines Case* (Washington, DC: WRI, 1992); Myers, op. cit. note 13; Costanza et al., op. cit. note 13; conversion to sustainable rural development in the Amazon from Fearnside, op. cit. note 12; Kumari, op. cit. note 19; Ruitenbeek, op. cit note 12.

130. Lynch and Talbott, op. cit. note 85.

131. Ibid.; Gadgil, op. cit. note 86; Poffenberger and McGean, op. cit. note 86; Broad, op. cit. note 3; Nicholas K. Menzies and Nancy L. Peluso, "Rights of Access to Upland Forest Resources in Southwest China," *Journal of World Forest Resource Management*, vol. 6, 1991; British Columbia from Burda et al., op. cit. note 97; and Ecotrust Canada, op. cit. note 108; Silva Forest Foundation, op. cit. note 80.

132. Tarasofsky, op. cit. note 106.

133. Treaty signatories from CBD Subsidiary Body for Scientific, Technical, and Technological Advice, <http://www.biodiv.org/sbstta.html>, viewed 20 October 1997.

134. United Nations, *Agenda 21: The United Nations Program of Action From Rio* (New York: U.N. Publications, 1992).

135. Tarasofsky, op. cit. note 106. As of mid-February 1998, the World Commission had not yet published its findings.

136. U.N., "United Nations Panel Proposes Action to Implement Earth Summit Forest Accords," press release (New York: 21 February 1997); "Plan on Forests Adopted by Ministers Leaves Question Open of Negotiating Treaty," *International Environment Reporter*, 9 July 1997; U.N. Department for Policy Coordination and Sustainable Development, "Report of the Open-ended Ad-hoc Intergovernmental Forum on Forests on Its First Session," advance unedited text, New York, 1–3 October 1997.

137. U.N., op. cit. note 136; Janet N. Abramovitz, "Another Convention Won't Save the Forests," *World Watch*, May/June 1997; "International Citizen Declaration Against a Global Forest Convention," released by various nongovernmental organizations in New York City at the fourth meeting of the Intergovernmental Panel on Forests, 10 February 1997; Government of Canada, "Canada Supports an International Forests Convention," press release, <http://www.nrcan.gc.ca/cfs/proj/ppiab/for-conv_e. html>, viewed 19 August 1997.

138. Seth Dunn, "After Kyoto: A Climate Treaty with No Teeth?" *World Watch*, March/April 1998; United Nations Framework Convention on Climate Change (UNFCC), "Kyoto Protocol to the United Nations Framework Convention on Climate Change," FCCC/CP/L.7/Add.1, 10 December 1997.

139. Opportunities for international cooperation include MERCOSUR, NAFTA, the World Trade Organization, the Central American Forest Agreement, RAMSAR, and Asia-Pacific Economic Cooperation, under negotiation; Tarasofsky, op. cit. note 106.

140. Dudley op. cit. note 2.

141. Ibid.

142. Ibid.; G-7 funding for Brazil from Carlos Sergio Figueiredo Tautz, "The Asian Invasion: Asian Multinationals come to the Amazon," *Multinational Monitor,* September 1997; Kumari op. cit. note 19; GEF from Hilary F. French, *Partnership for the Planet: An Environmental Agenda for the United Nations,* Worldwatch Paper 126 (Washington, DC: Worldwatch Institute, July 1995), and Stanley W. Burgiel and Sheldon Cohen, "The Global Environmental Facility From Rio to New Delhi: A Guide for NGOs" (Gland, Switzerland, and Cambridge, U.K.: IUCN-World Conservation Union, 1997); GEF funding as of June 1997 from "Project Implementation Review of the Global Environment Facility 1997," <http://www.gefweb. org/MONITOR/>, viewed 5 March 1998.

143. Hilary F. French, *Investing in the Future: Harnessing Private Capital Flows for Environmentally Sustainable Development,* Worldwatch Paper 139 (Washington, DC: Worldwatch Institute, February 1998); Bardacke, op. cit. note 93; Sullivan, op. cit. note 120; Cambodia from "Consultative Group Meeting on Cambodia, Paris, 1–2 July 1997," Global Witness, <http://www.oneworld.org/globalwitness>, viewed 23 September 1997; World Bank Forest Strategy Policy and Review, <http://www-esd. worldbank.org/forestry>, viewed 24 February 1998.

144. IMF, "IMF approves stand-by credit for Indonesia," press release no. 97/50 (Washington, DC: 5 November 1997); IMF, "Statement by the Managing Director on the IMF Program with Indonesia," News Brief No. 98/2, (Washington, DC: 15 January 1998); Paul Blustein and Sandra Sugawara, "Rescue Plan for Indonesia in Jeopardy," *Washington Post,* 7 January 1998; Dean Yates, "Fresh Reforms Hit Suharto Family Kin: Markets Wary," Reuters, 15 January 1998; Philip Shenon, "For the First Family of Indonesia, an Empire Now in Jeopardy," *New York Times,* 16 January 1998; Sander Thoenes, "Indonesian Wood Cartel Resists IMF Reforms," *Financial Times,* 13 February 1998.

Worldwatch Papers

No. of Copies

Worldwatch Papers by Janet N. Abramovitz

_____**Total copies (transfer number to order form on next page)**

PUBLICATION ORDER FORM

_____ *State of the World:* **$13.95**
The annual book used by journalists, activists, scholars, and policymakers worldwide to get a clear picture of the environmental problems we face.

_____ **Worldwatch Library: $30.00 (international subscribers $45)**
Receive *State of the World* and all six Worldwatch Papers as they are released during the calendar year.

_____ *Vital Signs:* **$12.00**
The book of trends that are shaping our future in easy to read graph and table format, with a brief commentary on each trend.

_____ **WORLD WATCH magazine subscription: $20.00 (international airmail $35.00)**
Stay abreast of global environmental trends and issues with our award-winning, eminently readable bimonthly magazine.

_____ **Worldwatch Database Disk Subscription: $89.00**
Contains global agricultural, energy, economic, environmental, social, and military indicators from all current Worldwatch publications including this Paper. Includes a mid-year update, and *Vital Signs* and *State of the World* as they are published. Can be used with Lotus 1-2-3, Quattro Pro, Excel, SuperCalc and many other spreadsheets.
Check one: _____ **IBM-compatible** _____ **Macintosh**

_____ **Worldwatch Papers—See list on previous page**
Single copy: $5.00 • 2–5: $4.00 ea. • 6–20: $3.00 ea. • 21 or more: $2.00 ea. (Call Vice President for Communications at (202) 452-1999 for discounts on larger orders.)

<u>$4.00*</u> Shipping and Handling *($8.00 outside North America)*
minimum charge for S&H; call (800) 555-2028 for bulk order S&H

_____ **TOTAL** (U.S. dollars only)

Make check payable to Worldwatch Institute

1776 Massachusetts Ave., NW, Washington, DC 20036-1904 USA

Enclosed is my check or purchase order for U.S. $_____

☐ AMEX ☐ VISA ☐ MasterCard _____
Card Number   Expiration Date

signature

name   **daytime phone #**

address

city   **state**   **zip/country**

phone: (202) 452-1999   fax: (202) 296-7365   e-mail: wwpub@worldwatch.org
website: www.worldwatch.org

Wish to make a tax-deductible contribution? Contact Worldwatch to find out how your donation can help advance our work.

Date Due

OCT 21 2006			